INFORMATION *and* TELECOMMUNICATION TECHNOLOGIES

THE IMPACT *of* THEIR ADOPTION *on* SMALL *and* MEDIUM-SIZED ENTERPRISES

ÉLISABETH LEFEBVRE
LOUIS A. LEFEBVRE

INTERNATIONAL DEVELOPMENT RESEARCH CENTRE
Ottawa • Cairo • Dakar • Johannesburg • Montevideo
Nairobi • New Delhi • Singapore

Published by the International Development Research Centre
PO Box 8500, Ottawa, ON, Canada K1G 3H9

Legal Deposit: 3rd quarter 1996
National Library of Canada

A microfiche edition is available

To order, please contact IDRC Books.
        Mail: PO Box 8500, Ottawa, ON, Canada K1G 3H9
        Phone: (613) 236-6163 ext. 2087
        E-mail: order@idrc.ca

Our catalogue may be consulted online.
        Gopher: gopher.idrc.ca
        World Wide Web: http://www.idrc.ca

This book may be consulted online at http://www.idrc.ca/books/focus.html

ISBN 0-88936-807-4

TABLE OF CONTENTS

FOREWORD

The increasing availability, use, and globalization of information and telecommunication technologies (ITs) have begun to create relationships between their application and the competitiveness and productivity of industry. What these relationships are and how the positive ones can be stimulated are important issues in the development of national public policies, particularly those relating to ITs.

Canada's International Development Research Centre (IDRC) has been working on areas related to this question over the past 4 years. In April and December 1993, IDRC brought together researchers and practitioners in Montevideo, Uruguay, to examine these issues, both generally and in relation to Latin America and the Caribbean. It was agreed during these consultations that the focus of further research should be ITs and their potential for stimulating competitiveness and productivity in small and medium-sized enterprises (SMEs).

It has since become clear that is not possible to define and develop effective IT policies without clearly understanding the factors that have led to the adoption of ITs and the characteristics of the firms where they have been adopted. Also, as these technologies are constantly and rapidly changing, the definition of policies needs to be informed by analyzing the application of specific technologies and their adoption in SMEs operating under certain conditions.

In this book, Élisabeth and Louis Lefebvre respond to these challenges. Through an analysis of the literature and the various methodological models, they identify the factors affecting the adoption of ITs, the decision-making process for their adoption by SMEs, and the impacts of such adoption. The result is a publication that provides a broad view of the state of relevant research and will be of immediate application to researchers worldwide who are involved in IT policy.

This book greatly benefited from an active discussion on the listserv established for communication among project participants, other researchers, and IDRC. This discussion enabled the authors to revise and extend some sections and to produce this final publication. I would like to thank all the participants in this process, particularly Carlos Correa and Otilia Vainstock of the University of

Buenos Aires, Roberto Hidalgo and José Lanusse of the Instituto de Investigaciones Socio-Económicas y Tecnológicas, Lucie Déschênes of the Centre for Information Technology Innovation, and Francisco Gatto of the Economic Commission for Latin America and the Caribbean. Their active participation in the discussions enabled the resulting publication to be more relevant to the practical question of establishing IT polices for SMEs.

Fay Durrant
Senior Program Specialist
IDRC

ACKNOWLEDGMENTS

The authors wish to thank Fay Durrant from the International Development Research Centre (IDRC) and Lucie Déschênes from the Centre for Information Technology Innovation (CITI) for their support and insightful comments. Many thanks go to Louise Bisson for her patience and meticulousness in typing and formatting this document (and its earlier versions) and to Thierry Vasseur and Mario Bourgault for their valuable assistance in the literature search. Bev Chataway from IDRC graciously helped us in our search for additional references, which proved to be very helpful. Comments and suggestions from all the participants "at" the electronic conference on the Internet, organized by IDRC, led to an improved final version of the document.

This report was financially supported by IDRC and CITI, but the opinions expressed are those of the authors.

INTRODUCTION

Information and telecommunication technologies (ITs) have long been identified as key factors in international competitiveness and have even radically modified the basis of competition. The dramatic influence of ITs will continue to determine the competitive posture of virtually all businesses in most countries of the world. However, the rate of IT adoption differs from country to country and, in a given country, from firm to firm. The delay observed in IT adoption could in certain sectors place the very survival of some firms in jeopardy. The adoption of ITs is thus considered a crucial strategic issue.

Context

Because of their potential impacts on the productivity and competitiveness of firms, ITs are of special interest to policymakers. How can we promote and facilitate the introduction and implementation of ITs? How can we accelerate IT diffusion in the various sectors of economic activity? How can we assess the impacts of ITs? This document will provide some answers to these questions in the specific context of small and medium-size enterprises (SMEs). Special attention will be paid to a few key sectors — the food and beverage industry, textiles, the garment and leather industry, chemicals and chemical products, metal products, machinery and equipment, the lumber industry, the electrical and electronics industries, the plastics industry, and the services sector (for example, accounting and engineering services) — and to experiences in the Organization for Economic Co-operation and Development (OECD) countries.

This document is expected to provide policymakers with some useful information that will help them define guidelines for the development of SMEs.

Objectives

The overall objective of the study was to identify in the current literature (1985 to date) appropriate theoretical support and methodological models for measuring the impact of IT on the productivity and competitiveness of SMEs in Latin America, the Caribbean region, and Canada. More specifically, this document has the following objectives:

- To identify and analyze factors affecting the adoption of ITs;
- To determine the characteristics of the decision-making process (that is, the way in which the decision to adopt ITs is made) that promote or hamper IT adoption; and
- To evaluate the impacts of IT adoption.

Organization of the document

In Chapter 1, ITs will be defined and classified according to different models, and the rate of diffusion of ITs in various countries will be examined. In Chapter 2, internal and external factors affecting the adoption of ITs will be analyzed. The characteristics of the decision-making process as a prime adoption factor will be discussed in Chapter 3. Chapter 4 will investigate the impact of ITs on productivity, key competitive dimensions, performance, and work and employment. The conclusion will provide a brief synthesis and some comments on methodological issues affecting the future design of interview guides or questionnaires.

To be as pragmatic as possible, each of the four chapters will present models whenever appropriate and conclude with some proposed operational measures, which could be used later in an empirical study.

The literature for this document was extremely rich and diverse because it came from various fields of research. To meet the objectives set out above, it was necessary to be both concise and exhaustive. The literature was extensive: close to 400 references are listed in the Bibliography, and these are organized to correspond to the chapters of the document and the four sections of Chapter 4.

CHAPTER 1
TYPOLOGIES AND RATES OF DIFFUSION

The obvious starting point for this chapter is an accurate definition of ITs. However, this is a rather confusing issue because numerous typologies exist. Furthermore, the specific characteristics, or attributes, of ITs (see "Characteristics of IT Applications: Primary Versus Secondary Attributes") have to take into account the context, in particular the different countries and different economic sectors, where the rates of IT diffusion vary (see "Diffusion of IT Applications in Different Countries and Different Sectors of Economic Activity").

A measure for assessing the level of IT adoption will be proposed in "A Proposed Measurement for Assessing the Level of Adoption and the Rate of Diffusion of ITs."

Classification of ITs: points of convergence and divergence

Numerous IT typologies exist; they vary according to the point of view of the researchers and practitioners who formulated them.

Classification from a conceptual perspective

From a conceptual perspective, we will attempt to define broad categories of applications based on the role they play in organizations independently of their use in different functional areas. Often, when scanning the literature, one finds applications categorized as they relate to the groupings identified in Figure 1: on one hand, there are applications that support operations-type activities, such as transaction processing, process control, and office automation; on the other hand, we find traditional information reporting, decision-support, and executive information systems.

One important assumption of this study was that it is increasingly difficult to dissociate IT applications that support production or operations-type activities from those that support more traditional managerial activities (Allen and Morton 1994).

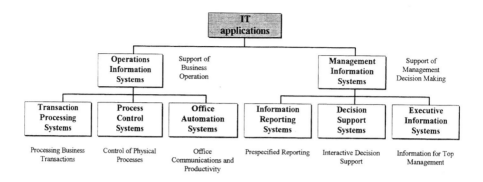

Figure 1. Classification of IT applications: a conceptual perspective. Source: O'Brien (1993).

Classification from a functional perspective

From a functional perspective, specific applications are not classified on the basis of broad types of information-processing activities (as identified in Figure 1) but are classified more in relation to the actual activities that must be carried out in the various functional areas of an organization. These applications are illustrated in Table 1.

The applications presented in Table 1 are usually found in large firms; in fact, some of them, such as the ones described in the category of human-resource management, would rarely apply in the context of smaller firms. However, because of the availability of easily accessible software packages, most of these applications, including the more sophisticated accounting applications, are becoming common in SMEs.

Classification from a technological perspective

The next perspective is the technological. The focus is on electronic data interchange (EDI), expert systems, and user computing systems or machines, regardless of the role they play in the organization or the specific activities they are used in. This technological focus will guide any IT-portfolio proposal and will further define the level of integration required among different technological applications. The issues of integration and interconnection are becoming increasingly important in most computer-based configurations, whether intrafirm or interfirm. To illustrate this perspective, let us consider a few examples.

The first example deals with an application that is becoming more and more common in manufacturing. The application uses computer-aided design (CAD) output to control machines used in manufacturing, that is, computer-aided

Table 1. Classification of IT applications: a functional perspective.

Production and operations	Marketing	Finance	Accounting	Human-resource management
CAM	Market research	Capital budgeting	Billing and accounts receivable	Personnel record-keeping
MRP I	Sales forecasting	Cash management	Payroll	Labour analysis
Inventory control	Advertising and promotion	Credit management	Accounts payable	Employee-skills inventory
Purchasing and receiving	Sales-order processing	Portfolio management	General ledger	Compensation analysis
Process control	Sales management	Financial forecasting	Fixed-asset	Training and development analysis
CAE	Product management	Financing-requirements analysis	accounting	Personnel-requirements forecasting
Robotics	Marketing management	Financial-performance analysis	Cost accounting	
			Tax accounting	
			Budgeting	
			Auditing	

Source: O'Brien (1990).
Note: CAE, computer-aided engineering; CAM, computer-aided manufacturing; MRP I, material-requirements planning.

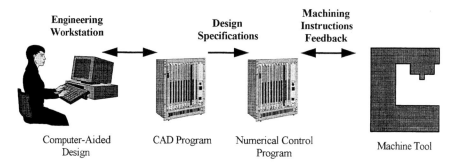

Figure 2. CAD–CAM in a manufacturing setting. Source: O'Brien (1990).

manufacturing (CAM), and is referred to as CAD–CAM. Figure 2 illustrates computers being used in CAD–CAM from the design stages, where the parts, components, or products are actually designed and defined to precise specifications, right through to the actual machining instructions for the computerized numerically controlled (CNC) machine tools.

The next example concerns EDI, which can be viewed as a three-party — the vendor (usually the supplier), the customer, and the buying and selling company (usually one large company) — electronic service (Figure 3) but still represents basically functional applications (billing, accounts receivable, and inventory management and control) supporting daily business operations.

Interfirm connections, such as those shown in Figure 3, are not only becoming more common today but are also becoming necessary for any firm that supplies the large companies that run on just-in-time (JIT) manufacturing cycles. These first-tier and second-tier suppliers, which are often SMEs, must be able to quickly provide parts and components for the larger manufacturer's inventory. This can only be done efficiently through IT networks, which monitor not only the flow of goods but also financial information regarding these goods or services.

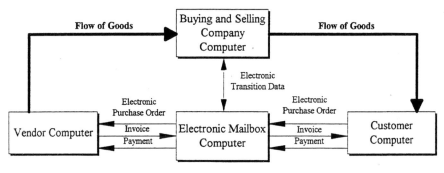

Figure 3. Electronic data exchange as a third-party service. Source: Adapted from Burch (1989).

These are merely illustrations of how technological applications can be integrated in and between firms. However, they point to the importance of technology in supporting and enhancing a firm's operations and contributing to the competitiveness of its products and services.

Classification from a sectorial perspective

The fact that different authors use different typologies of IT applications is not merely due to different objectives but is also due to the effect of the industrial sector of activity. Table 2 presents some of the more current basic and advanced applications found in different industry segments. Overall, the table provides a good indication of the industry focus and, therefore, of what IT contributions are expected.

We have presented four different perspectives (conceptual, functional, technological, and sectorial) offering different ways of classifying IT applications.[1] However, to properly assess the level and rate of use of IT applications (see "A Proposed Measurement for Assessing the Level of Adoption and the Rate of Diffusion of ITs"), these perspectives should be combined.

Characteristics of IT applications: primary versus secondary attributes

It is clear that the nature of IT applications is difficult to grasp, and viewpoints sometimes seem contradictory and fragmentary. Such difficulties arise from our failure to make a distinction between primary and secondary attributes (or characteristics) of IT applications. Primary attributes relate to the object (IT application) and are independent of the subject (the organization), whereas secondary attributes vary according to the subject's perception of the object.

Secondary attributes are more likely than primary attributes to shed light on the adoption factors and impacts of IT applications. Let us consider an example. All IT applications share a common primary attribute for all types of organizations in all industries: they are all computer-based applications. However, some of these applications may be considered either radical or incremental, depending on the organizational and industrial contexts. In a small firm, a specific IT application may be seen as radical because it requires a large portion of the available financial resources and heavy involvement on the part of qualified technical employees and implies major organizational changes. However, the same application may be considered incremental in a large firm.

---

[1] Other classifications exist: for example, the classical distinction between hard and soft technologies made by Swamidass (1994) and presented in Appendix A is interesting because the difficulties and impacts of adopting and implementing soft technologies are often underestimated.

Table 2. Information-systems applications by industry category.

Industry segment	Basic applications	Advanced applications
Manufacturing	Production accounting Production planning Purchasing and receiving Process control Inventory control	CAM CAE Process control Inventory control Numerical control Robotics
Business and personal services	Service-bureau functions Tax preparation Accounting Client records	Econometric models Time-sharing Engineering analysis Financial planning
Banking and finance	Demand deposit accounting Cheque processing Proof and transit operations Cost accounting	On-line savings Electronic funds transfer Portfolio analysis Cash-flow analysis
Insurance	Premium accounting Customer billing External reports Reserve calculation	Actuarial analysis Investment analysis Policy approval Cash-flow analysis
Utilities	Customer billing Accounting Meter reading Inventory control	Rate analysis Line and generator loading Operational simulation Financial models
Distribution	Order processing Inventory control Purchasing Warehouse control	Vehicle scheduling EDI Forecasting Store-site selection
Transportation	Rate calculation Vehicle maintenance Cost analysis Accounting	Traffic-pattern analysis Automatic rating Tariff analysis Reservation systems
Health care	Patient billing Inventory accounting Health-care statistics Patient history	Lab–operation scheduling Nurses' station automation Patient monitoring Computerized diagnostics
Retailing	Customer billing Sales analysis Accounting Inventory reporting	Point-of-sale systems Sales forecasting Merchandising EDI
Printing and publishing	Circulation Classified ads Accounting Payroll	Automated typesetting Desktop publishing Media analysis Page layout

Source: O'Brien (1990).
Note: CAE, computer-aided engineering; CAM, computer-aided manufacturing; EDI, electronic data interchange.

The same line of reasoning applies to the size of the firm, to the industrial sector (some sectors are more technologically sophisticated than others), and to country-specific contextual factors (such as the presence of a more appropriate national technological infrastructure). Differences between small and large firms, industries, and countries exist, as illustrated in the next section.

Diffusion of IT applications in different countries and different sectors of economic activity

This section will present the actual use of some IT applications in different countries[2] and, in specific countries, their rate of diffusion. We will also examine the rate of penetration in different sectors of activity and then take a closer look at SMEs.

Use of IT applications in different countries

The need for detailed, comparable, and up-to-date information about the rate of diffusion of IT applications in different countries is strong but still remains partially unmet, although OECD has made many efforts to promote international comparability in technology use (see, for example, OECD 1990, 1992). The most comprehensive effort to gather and summarize this information was made by Northcott and Vickery (1993). They extensively reviewed all surveys done in OECD countries since 1980 and tried to analyze the overlap between the various issues covered in these national surveys (Table 3).

As shown in Table 3, the overlap among the surveys carried out in 13 countries is greatest for the use of computer-based advanced manufacturing technologies (AMTs), which results in the internationally comparable data[3] presented in Table 4.

Table 4 indicates that in the late 1980s Japan had a clear lead in robots, flexible manufacturing systems (FMS), and automated storage and retrieval systems (AS–RS). The United States and Canada seemed to lag behind Japan in

---

[2]   The rate of penetration of ITs in different countries can be assessed by the level of IT investments expressed as a percentage of gross domestic product (GDP), which is a rather rough proxy. Small countries (having fewer than 10 million people) do invest in ITs: 2.7% of GDP for New Zealand; 2.2%, for Singapore; 1.5%, for Denmark, Hong Kong, and Norway; 1.3%, for Sweden; 1.1%, for Finland; and 2.8%, for the United States (Dedrick et al. 1995).

[3]   Three surveys — the 1988 survey by the US Bureau of Census (1989), the second Canadian survey by Statistics Canada (1989), and the 1988 Australian survey by the Australian Bureau of Statistics (ABS 1989, 1990) — were the most extensive because they share data on the use and planned use of 17 technologies. The Australian survey included advanced cutting technologies, apart from lasers and different types of robots (pick-and-place robots, arc-welding and spot-welding robots, and those used for assembly or finishing).

Table 3. Issues covered by surveys on microelectronics and

	Australia	Austria	Canada	Denmark	Finland
Use in products					
Actual		X		X	X
Planned		X		X	X
Use in manufacturing processes					
Actual	X	X	X	X	X
Planned	X	X	X	X	X
Technology source	X	X		X	X
Motives for nonuse		X	X		X
CAD	X	X	X	X	X
CNC	X	X	X	X	X
Robots	X	X	X	X	X
FMS	X	X	X	X	X
CAM	X	X	X	X	
Inspection and testing	X	X	X	X	X
AS–RS	X	X	X	X	X
Communication and control	X	X	X	X	X
Motives for use					
In products					X
In processes	X	X		X	X
Competencies and impacts					
Number of engineers				X	X
Training programs	X	X	X	X	X
Impact on employment		X			X
Level of investment	X				
Government support			X		
Problems for use					
In products					X
In processes		X		X	X

Source: Northcott and Vickery (1993).
Note: AS–RS, automated storage and retrieval systems; CAD, computer-aided design; CAM, computer-aided manufacturing; CNC, computerized numerically controlled (machines); FMS, flexible manufacturing systems.

advanced-technology use in OECD countries.

France	Germany	Japan	Netherlands	New Zealand	Sweden	UK	USA
X	X		X	X	X	X	
X	X		X	X	X	X	
X	X	X	X	X	X	X	X
X	X	X	X	X	X	X	X
X	X					X	X
					X	X	X
X	X		X	X	X	X	X
X	X	X	X	X	X	X	X
X	X	X	X	X	X	X	X
		X		X	X		X
			X		X		X
X	X	X	X	X	X		X
X	X		X	X	X	X	X
X	X	X	X	X	X	X	X
X			X			X	X
				X		X	X
X	X		X	X	X	X	X
X	X	X			X	X	
X	X		X			X	
			X				
					X	X	
X	X	X		X	X	X	X
X	X			X	X	X	

Table 4. Use of advanced manufacturing technologies in OECD countries by firm size.

Computer-based production technologies	Use by country (%)											
	Australia (1988)		Canada (1989)		Japan (1988)		Sweden (1986)		UK (1987)		USA (1988)	
	10–199	≥200	20–99	≥500	<300	≥300	5–19	≥500	1–199	≥200	20–499	≥500
CAD–CAE	8.3	39.0	27.0	75.0	39.1	75.2	7.7	57.7	5.6	42.0	36.3	82.6
NC–CNC machines	11.8	34.8	21.0	65.0	57.4	79.4	—	—	6.2	33.0	39.6	69.8
FMC–FMS	0.9	5.1	7.0	31.0	39.4	67.4	—	—	—	—	9.1	35.9
Pick-and-place robots	1.5	14.7	3.0	30.0	22.6	62.2	—	—	1.4	15.0	5.5	43.3
Assembling robots	0.3	4.3	—	—	8.3	41.4	—	—	0.5	0.9	3.9	35.0
AGVS	0.5	2.7	2.0	12.0	—	—	—	—	—	—	0.8	13.1
AS–RS	1.3	11.7	1.0	11.0	10.9	44.9	1.6	26.7	—	—	1.9	24.4

Source: Northcott and Vickery (1993).
Note: AGVS, automated guided-vehicle systems; CAD, computer-assisted design; CAE, computer-assisted engineering; CNC, computerized numerically controlled; FMC, flexible manufacturing cell; FMS, flexible manufacturing systems; NC, numerically controlled.

all technologies except computer-aided design and engineering (CAD–CAE) and numerically controlled (NC) machines. In all countries, the use of these advanced technologies was much lower in SMEs than in their larger counterparts. The data shown in Table 4 should be interpreted cautiously because the surveys were not carried out at the same time (1986–89) and because the firms' sizes differed from country to country. More recently, Japan's lead in industrial-robot use was again confirmed (Figure 4).

Rate of diffusion of IT applications in specific countries

Longitudinal surveys are necessary to assess the rate of diffusion. The most comprehensive survey carried out between two points in time was certainly the one published in February 1995 by Statistics Canada (1995).

Table 5 shows the advances in technology use in Canadian manufacturing firms between 1989 and 1993. The highest growth rates were for CAD–CAE (19%) and the use of local-area networks (LANs) for technical data (14%). Automated handling systems (AS–RS and automated guided-vehicle systems [AGVS]) and most applications related to manufacturing and assembly had the lowest growth rates in Canadian manufacturing firms. Were these rates of penetration similar in other countries?

The longitudinal survey carried out by Swamidass (1994) and published by the National Association of Manufacturers (United States) provided partial answers to that question. Swamidass introduced a fundamental distinction not found in the

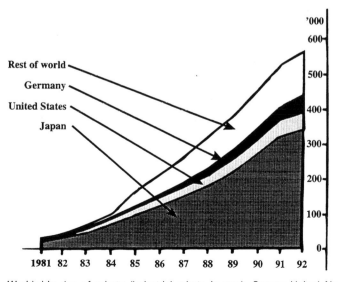

Figure 4. Worldwide rise of robots (industrial robots in use). Source: United Nations, in *The Economist* (March 1994).

Table 5. Growth in technology use between 1989 and 1993 (shipment weighted) for Canadian manufacturing firms.

Computer-based production applications	Use (% of shipments)			
	1989 survey		1993 survey (modified) in 1993	Change (%)
	In 1989	Projected		
Design and engineering				
CAD–CAE	49	58	68	19
CAD–CAM	20	26	24	4
Digital representation of CAD output used in procurement activities	13	20	20	7
Fabrication and assembly				
FMC–FMS	21	25	23	2
NC–CNC machines	30	32	31	1
Materials-working lasers	9	15	9	0
Pick-and-place robots	15	18	23	8
Other robots	16	19	15	−1
Automated handling systems				
AS–RS	15	23	15	0
AGVS	9	17	10	1
Inspection and communications				
Automated inspection equipment for inputs	31	39	38	7
Automated inspection equipment for final products	35	42	46	11
LAN for technical data	41	53	55	14
LAN for factory use	37	49	46	9
Intercompany computer network	35	47	39	4
Programable controllers	64	66	68	4
Computers used for control in factories	50	58	62	12
Manufacturing information systems				
MRP I	49	60	59	10
MRP II	33	48	42	9
Integration and control				
CIM	21	29	29	8
SCADA	34	43	43	9
Artificial intelligence– expert systems	7	20	12	5

Source: Statistics Canada (1995).
Note: AGVS, automated guided-vehicle systems; AS–RS, automated storage and retrieval systems; CAD, computer-aided design; CAD–CAM, CAD output used to control manufacturing machines; CAE, computer-aided engineering; CAM, computer-aided manufacturing; CIM, computer-integrated manufacturing; CNC, computerized numerically controlled; FMC, flexible manufacturing cell; FMS, flexible manufacturing systems; LAN, local-area network; MRP I, material-requirements planning; MRP II, manufacturing-resource planning; NC, numerically controlled; SCADA, supervisory control and data acquisition.

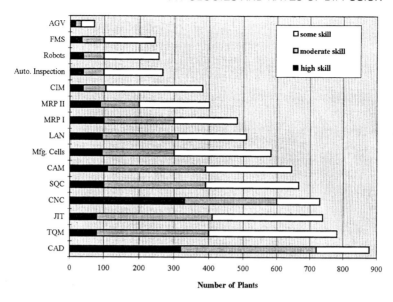

Figure 5. Technology use in US plants by skill level. Source: Swamidass (1994). Note: AGV, automated guided vehicle; CAD, computer-aided design; CAM, computer-aided manufacturing; CIM, computer-integrated manufacturing; CNC, computerized numerically controlled (machines); FMS, flexible manufacturing systems; JIT, just in time; LAN, local-area network; MRP I, material-requirements planning; MRP II, manufacturing-resource planning; SQC, statistical quality control; TQM, total quality management.

national surveys: the use of a particular application as such is not enough if we do not take into account the existing level of skills related to that application in each firm. Swamidass suggested that there are three types of users: highly skilled users, moderately skilled users, and users with some skills (Figure 5). In fact, as was the case for Canadian firms, CAD was the most widely used application, but only one-third of US manufacturing firms were highly skilled in its use. Although CNC was less frequently used, the percentage of skilled users of CNC was similar to that for CAD. In general, the levels of highly skilled use were fairly low for JIT, computer-integrated manufacturing (CIM), manufacturing-resource planning (MRP II), automated inspection, robots, FMS, and AGVs.

Plans to become highly skilled in technology use appeared to have changed between 1990 and 1994 among US manufacturers. Table 6 indicates that CAD and JIT were the most frequently emphasized in 1990, but in 1994, JIT moved ahead of CAD. The last column of Table 6 is quite interesting because it shows the shift in US manufacturers' interest: flexible manufacturing cells (FMCs) showed the largest rise in interest, whereas CAD showed a definite decline in interest, probably as a result of the near saturation in its use.

Table 6. Changes in plans to become highly skilled technology users in US manufacturing firms (among participants in both studies).

Technology	Second study (%)	First study (%)	Second–first difference (%)
FMC	39.3	32.2	+7.1
SQC	33.7	28.6	+5.1
JIT	48.6	43.6	+5.0
FMS	12.5	9.1	+3.4
MRP I	21.7	21.3	+0.4
CAD	45.3	50.9	−5.6
Robots	8.3	14.0	−5.7
MRP II	20.3	29.6	−9.3
CAM	30.0	39.4	−9.4

Source: Swamidass (1994).
Note: CAD, computer-aided design; CAM, computer-aided manufacturing; FMC, flexible manufacturing cell; FMS, flexible manufacturing systems; JIT, just in time; MRP I, material-requirements planning; MRP II, manufacturing-resource planning; SQC, statistical quality control.

Use of IT applications in specific sectors

In 1995, the use of IT applications was found to vary from industry to industry. Table 7 shows that three industries — electrical and electronic products, primary metals, and transportation equipment — were the chief adopters of computer-based production applications in Canada (Statistics Canada 1995), whereas lumber, rubber and plastics, textiles and clothing, and furniture and fixtures showed the lowest penetration rates. The industries of particular interest to this study are indicated. If we look at firms using at least one technology, we find that the same industries as those shown in Table 7 were leaders and laggards (Figure 6).

The services sector does not seem to attract the same attention from national agencies as the manufacturing sector. The notable exception was a report by the Economic Council of Canada (McFetridge 1992).

The use of ITs in the services sector is presented in Table 8. Personal computers and facsimile machines were in 1992 the most widely used, with close to 90% of establishments using this equipment. Some office networking applications, such as voice mail or video conferencing, were not adopted to any great extent.

Table 7. Functional technology use by industry (shipment weighted).

Industry	Use (% of shipments)					
	Design and engineering	Fabrication and assembly	Automated materials handling	Inspection and communications	Manufacturing information systems	Integration and control
Electrical and electronic products [a]	86.4	77.4	52.7	84.8	78.3	61.6
Primary metals [a]	90.3	70.6	12.2	91.8	70.6	64.1
Transportation equipment [a]	74.8	77.1	10.7	89.4	69.4	45.5
Paper and allied products	76.0	30.4	28.9	79.5	26.5	38.5
Petroleum and chemical products [a]	66.8	24.7	7.9	80.6	65.1	62.5
Other manufacturing industries	69.9	64.6	6.3	70.0	55.6	31.5
Machinery [a]	75.1	64.7	11.9	61.4	43.4	24.2
Food processing [a]	48.1	26.1	17.5	70.0	53.5	31.6
Nonmetallic mineral products [a]	25.7	45.2	31.9	58.6	38.5	44.0
Fabricated metal products [a]	44.5	33.6	0.5	30.5	26.3	4.2
Lumber [a]	37.1	34.9	9.8	55.6	6.9	21.3
Rubber and plastics [a]	37.7	34.9	9.7	48.9	26.6	16.4
Textiles and clothing [a]	44.6	26.3	14.5	45.0	23.1	34.4
Furniture and fixtures [a]	44.1	39.7	2.4	25.4	37.1	13.6
Printing and publishing	31.3	16.2	4.3	50.9	23.5	21.2

Source: Statistics Canada (1995).

[a] Industries of particular interest to this study.

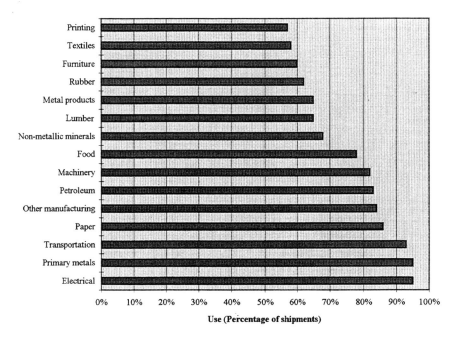

Figure 6. Adoption rates of advanced technologies by industry in Canada (percentage of firms using at least one technology). Source: Statistics Canada (1995).

According to the same study (McFetridge 1992), the following specific service sectors showed the highest adoption rates:

- Communications industry;
- Wholesale trade;
- Finance and insurance industry; and
- Business-service industries.

The lowest rates of adoption (McFetridge 1992) were found in the following sectors:

- Accommodation industry;
- Food and beverage industry; and
- Retail trade.

Some IT applications, such as transportation systems, computerized reservation systems, and property-management systems, were industry specific in 1990; this is illustrated in Figure 7.

Table 8. Services-sector establishments using or planning to use selected technologies in Canada.

Technology	Using (%)	Planning to use within 3 years (%)
Office automation		
Personal computers	89	3
On-line terminals	76	4
Minicomputers	54	4
Mainframe computers	41	2
Office networking		
Facsimile	89	3
LAN	40	17
Telex	36	0
Electronic mail (private)	30	14
WAN	29	10
External databases	22	8
Mobile data communications	11	5
Electronic mail (public)	10	9
Voice mail	6	7
Satellite data distribution	3	5
Video conferencing	2	5
Design support systems		
DTP	30	15
CAD	14	5
CAE	6	4
CASE	6	8
Inventory–sales systems		
Computerized inventory control	56	12
Computerized order entry	50	9
Point-of-sale terminals	22	8
EDI	19	16
Electronic scanning systems	15	14
Automated retrieval systems	10	4

Source: McFetridge (1992).
Note: CAD, computer-aided design; CAE, computer-aided engineering; CASE, computer-aided software engineering; DTP, desktop publishing; EDI, electronic data interchange; LAN, local-area network; WAN, wide-area network.

Logically enough, transportation systems, such as fleet-management and freight-analysis systems, were found primarily in the transportation industry and to a much lesser extent in the wholesale-trade sector (Figure 7). Computerized reservation systems were most widely used by hotels and restaurants (accommodation and food industry), whereas property-management systems were mostly adopted by real-estate operators.

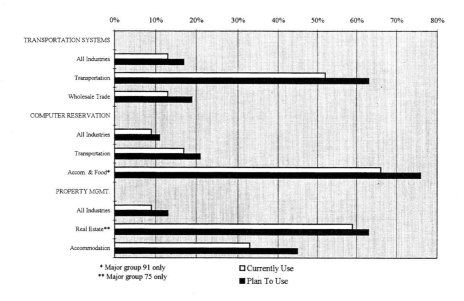

Figure 7. Industry-specific IT applications in the Canadian services sector. Source: Industry, Science and Technology Canada (1990).

Rate of diffusion of IT applications in SMEs

All national surveys pointed to the same phenomenon: larger establishments were making greater use of IT applications in both the manufacturing sector and the services sector.

The recent survey carried out by Statistics Canada (1995) demonstrated a constant progression in the use of ITs from artisanal firms (0–19 employees), to small firms (20–99 employees), to medium-sized firms (100–499 employees), to large firms (≥500 employees), both in the case of broad functional advanced-manufacturing application (Table 9) and in the case of each specific manufacturing application (Table 10).

Not surprisingly, Swamidass (1994) found that AMTs (Figure 8) were more common in large US plants (with more than 100 employees) than in small plants (with fewer than 100 employees). In the case of some technologies (for example, FMS or AGVS), the percentage of large plants using these technologies was much greater than that of small plants. In the case of CAD, small plants were catching up with their larger counterparts (75.5% versus 94.8%). Affordability and internal expertise and know-how no doubt explained these differences.

In the services sector, large organizations (with more than 200 employees) also made greater use of ITs in 1990 (Figure 9).

Table 9. Functional technology use by employment size (shipment weighted) for Canadian manufacturing firms.

Technology	Number of employees			
	0–19	20–99	100–499	≥500
	Use (% of shipments)			
Design and engineering	30.7	39.2	57.7	81.8
Fabrication and assembly	19.6	19.6	37.0	67.4
Automated materials handling	8.9	4.4	12.3	17.5
Inspection and communications	34.9	41.2	73.3	93.7
Manufacturing information systems	22.3	25.1	40.9	81.5
Integration and control	23.3	22.3	34.2	56.4

Source: Statistics Canada (1995).

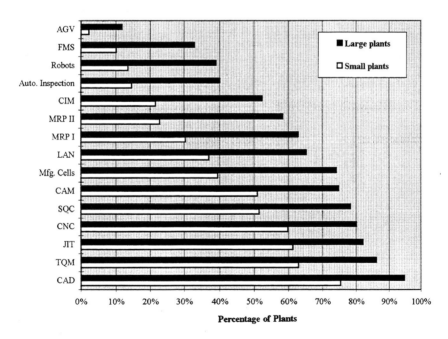

Figure 8. Technology use in small and large US plants. Source: Swamidass (1994). Note: AGV, automated guided vehicle; CAD, computer-aided design; CAM, computer-aided manufacturing; CIM, computer-integrated manufacturing; CNC, computerized numerically controlled (machines); FMS, flexible manufacturing systems; JIT, just in time; LAN, local-area network; MRP I, material-requirements planning; MRP II, manufacturing-resource planning; SQC, statistical quality control; TQM, total quality management.

Table 10. Advanced technology use in 1993 by employment size (shipment weighted) for Canadian manufacturing firms.

Technology	Number of employees			
	0–19	20–99	100–499	≥500
	Use (% of shipments)			
Design and engineering				
CAD–CAE	28.0	35.8	55.8	81.2
CAD–CAM	8.7	10.9	13.2	27.6
Digital representation of CAD output used in procurement activities	5.1	6.0	12.2	22.4
Fabrication and assembly				
FMC–FMS	7.4	6.9	11.9	28.1
NC–CNC machines	14.1	13.1	21.4	35.8
Materials-working lasers	1.2	2.7	3.7	16.1
Pick-and-place robots	2.2	2.4	12.7	32.3
Other robots	2.3	2.6	10.6	27.9
Automated handling systems				
AS–RS	7.5	4.3	11.1	12.8
AGVS	4.6	0.6	3.0	8.5
Inspection and communications				
Automated inspection equipment for inputs	15.6	8.1	25.1	45.3
Automated inspection equipment for final products	16.5	11.9	29.4	59.2
LAN for technical data	22.4	16.0	39.2	71.1
LAN for factory use	22.8	11.8	34.4	57.4
Intercompany computer network	11.6	7.7	23.8	54.6
Programable controllers	22.5	22.6	57.0	78.3
Computers used for control in factories	24.4	24.4	49.3	71.0
Manufacturing information systems				
MRP I	22.1	23.1	39.2	73.8
MRP II	15.7	14.5	21.6	58.6
Integration and control				
CIM	16.0	15.7	14.8	29.8
SCADA	17.7	15.7	30.7	46.2
Artificial intelligence—expert systems	1.0	0.6	8.4	16.7

Source: Statistics Canada (1995).
Note: AGVS, automated guided-vehicle systems; AS–RS, automated storage and retrieval systems; CAD, computer-aided design; CAD–CAM, CAD output used to control manufacturing machines; CAE, computer-aided engineering; CAM, computer-aided manufacturing; CIM, computer-integrated manufacturing; CNC, computerized numerically controlled; FMC, flexible manufacturing cell; FMS, flexible manufacturing systems; LAN, local-area network; MRP I, material-requirements planning; MRP II, manufacturing-resource planning; NC, numerically controlled; SCADA, supervisory control and data acquisition.

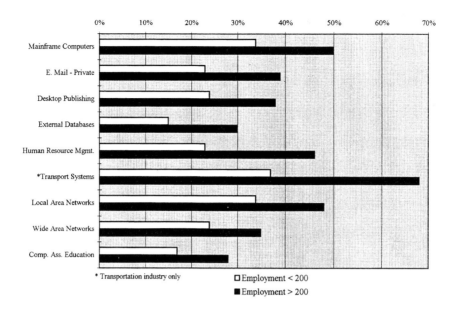

Figure 9. Technology use in the services sector by small and large Canadian firms. Source: Industry, Science and Technology Canada (1990).

Although IT applications were in 1994 more often present in large firms, small businesses were increasingly turning to state-of-the-art ITs, and the vast majority of small firms surveyed considered these technologies a critical or important factor (Figure 10). Improvements in computer hardware and software, along with falling prices, placed these IT applications within the grasp of more and more small firms.

Which applications were the most valued? Small US firms favoured the following in decreasing order: payroll, tax, and booking; order entry and billing;

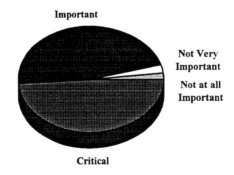

Figure 10. Importance of ITs in small US firms. Source: *Business Week* (1994).

Table 11. Computer-based applications in small US firms.

	Use (%)
Most-valued systems	
Payroll, tax, and booking	74
Order entry and billing	65
Financial analysis–cash management	54
Sales information	52
Least-valued systems	
Inventory control	36
Customer service	22
Distribution and logistics	17
Manufacturing	16

Source: *Business Week* (1994).

financial analysis–cash management; and sales information (Table 11). Distribution and logistics systems and manufacturing systems were the least-adopted IT applications among small firms but were still present in almost one out of five.

There is no doubt that ITs can now reach firms of every size and in every sector of the economy.

A proposed measurement for assessing the level of adoption and the rate of diffusion of ITs

We propose that the concept of IT be defined in terms of specific applications. Therefore, microcomputers or microprocessors integrated in the products and services offered by a particular firm will be excluded. For example, "smart" products, such as credit cards with chips or automobiles with microcomputers, will not be considered. This also suggests that the focus must be on the adoption of computer-based applications, such as inventory management, rather than on the possession of a microcomputer.

The concept of IT should also include both intrafirm and extrafirm applications:[4] the fusion of computers, information systems, and telecommunications is a fact, and applications like EDI should therefore be included.

---

[4]   These include networking applications in the firm (for example, the use of a LAN for technical data) and between firms (for example, EDI with customers or suppliers).

The proposed list of applications (Table 12) was derived from the list first established by Statistics Canada (1989), which was tested extensively in the specific context of small manufacturing firms (Lefebvre and Lefebvre 1992). This allowed us to eliminate many of the applications that do not apply to smaller firms that because of their size do not require such applications in the first place or cannot handle their technical sophistication. Note that some of these applications can only be found in very specialized manufacturing firms operating in industrial environments where product specifications and requirements are strict and well defined (for example, the electronics or aeronautics industries).

The list of applications devised by Statistics Canada (1989) is very well known and can easily be compared with those derived from the United States and Australia (Ducharme and Gault 1992). The list is also similar in many respects to the one used by Swamidass (1994). These other lists of applications exclude computer-based administrative applications, but we strongly believe that administrative applications should be included because they are crucial in both the services and the manufacturing sectors.

Moreover, in manufacturing firms, computer-based administrative and production applications are developed simultaneously, and synergistic effects are observed between the two types of application. It is therefore impossible to dissociate these two types of applications when trying to determine the relative influence of various adoption factors or to assess their impact. Applications such as CAD–CAM, computer-aided engineering (CAE), material-requirements planning (MRP I), MRP II, and FMS should be included because they all rely on information systems. Some authors have indeed stressed the importance of cross-functional information systems as an effective and efficient way to attain strategic objectives (Moad 1989; Wilder 1989), but this integration is difficult to achieve for most firms. From a practical point of view, sharing databases is still considered by chief executive officers (CEOs) a crucial issue facing organizations, although data sharing across application systems and departments is at the heart of cross-functional information systems. Going one step further, the integration of computer-based administrative applications and AMTs certainly results in numerous managerial complexities that seem to be much greater than the technical complexities, mainly because of the split between administrative employees (marketing, sales, accounting, or finance) and production employees (engineers, production managers, machinists, technicians, or specialized blue-collar workers).

Such integration is therefore complex, but this by no means precludes synergy between these technologies; this is true for two compelling reasons. First, AMTs are considered the "manufacturing subset of information technology," and

Table 12. List of IT applications.

Computer-based administrative applications [a]	Computer-based production applications [b]	Computers used for control in factories
Accounts payable—accounts receivable	Design and engineering	Manufacturing information systems
Inventory management and control	CAD and CAE	MRP I
Sales analysis	CAD–CAM	MRP II
Payroll	Digital representation of CAD output for procurement activities	Integration and control
Billing	Fabrication and assembly	CIM
Cost accounting	FMC or FMS	SCADA
Operations management and production scheduling	NC and CNC machines	Artificial intelligence—expert systems
Word processing	Materials-working lasers	
Electronic mail—electronic filing	Pick-and-place robots	
Teleconferencing—video conferencing	Other robots	
Expert systems	Automated materials handling	
EDI with customers	AS–RS	
EDI with suppliers	AGVS	
	Inspection and communications	
	Automated inspection equipment for inputs	
	Automated inspection equipment for final products	
	LAN for technical data	
	LAN for factory use	
	Intercompany computer network	
	Programable controllers	

Note: AGVS, automated guided-vehicle systems; AS–RS, automated storage and retrieval systems; CAD, computer-aided design; CAD–CAM, CAD output used to control manufacturing machines; CAE, computer-aided engineering; CAM, computer-aided manufacturing; CIM, computer-integrated manufacturing; CNC, computerized numerically controlled; FMC, flexible manufacturing cell; FMS, flexible manufacturing systems; LAN, local area network; MRP I, materials-requirements planning; MRP II, manufacturing-resource planning; NC, numerically controlled; SCADA, supervisory control and data acquisition.
[a] Extensively used in several studies (for example, Lefebvre and Lefebvre 1992).
[b] Adapted from a typology produced by Statistics Canada (1995). Definitions are offered in Appendix B.

FMS are considered one of the "most important industrial applications of information technology" (Mansfield 1993). Although traditional manufacturing systems use paper-based information systems, the AMTs both thrive and depend on highly sophisticated IT.

Second, as AMTs become more and more integrated — moving from hybrid cell to assembly cell, to machining cell, to FMS, and finally to CIM — more and more interdepencies are created and higher levels of integration of ITs are required (Gerwin and Kolodny 1992). CIM implies the integration of production with engineering and business functions (Hsu and Skevington 1987; Thacker 1989), and thus a logical consequence is the integration of computer-based administrative applications and AMTs. We therefore argue that these two broad types of technologies cannot be dissociated, even in small firms where such integration is just starting.

Proposed measurement of the level of adoption of IT applications

The level of IT adoption is a composite measure that takes into account the number of applications adopted by a firm and a weighting attributed by a panel of experts, who rank each application according to its degree of radicalness–innovativeness[5] (see "Characteristics of IT Applications: Primary Versus Secondary Attributes"). The experts chosen must have considerable expertise in ITs and must also be very familiar with the context in which the firms operate.

This proposed measurement will allow an adequate comparison from country to country and from firm to firm and result in a score that is simply calculated, as follows (Lefebvre and Lefebvre 1992):

$$\sum_{i=1} i_j \times r_j$$

where $i_j = 0$ or 1, depending on the adoption of innovation j; and r_j is the degree of radicalness of innovation j as established by a panel of experts who ranked each innovation on seven-point Likert scales.

To assess the degree of penetration of IT applications, it is possible to estimate the percentages of employees who are highly skilled users, moderately skilled users, and users with some skills in each application.

Finally, one can attempt to determine the level of integration between these applications, which is rather difficult and requires on-site observations.

---

[5] The degree of radicalness differs for different organizational contexts (that is, larger versus smaller SMEs) and for different industrial contexts (that is, specific sectors of activity).

Proposed measurement of the rate of diffusion of IT applications

In the case of longitudinal studies (that is, assessing the level of adoption of IT at two different points in time), the rate of diffusion of these technologies can be easily estimated. When longitudinal studies are excluded, it will still be possible to estimate the rate of diffusion by simply asking about the use planned in 2 or 3 years. This will be a rough estimate because some plans may never be realized, but it will still provide some indication.

The advantages of the two proposed measurements are numerous: they can act as a benchmark among firms, industries, and countries; they have been extensively pretested in large and small firms; and they are simple and repeatable. Because the adoption of IT applications is context specific, some applications may be removed from the list presented in Table 12, or some industry-specific applications, such as the ones shown in Table 2, can be added. The proposed measurement is therefore easily adaptable to different contexts.

CHAPTER 2

FACTORS AFFECTING ADOPTION

Factors promoting or hampering the adoption of IT applications are numerous and have been a prime concern for many researchers and practitioners (see the bibliography). In this rich variety of literature, adoption factors were sometimes examined in the specific context of SMEs but can be more broadly considered as originating from inside or outside the organization. Figure 11 presents these two different levels of determinants, but both internal and external factors must be taken into account when trying to understand a firm's criteria for deciding about technology. The best case, obviously, is one in which it is possible to ultimately find congruence, or fit, between the factors internal to and those external to the firm.

Adoption factors are usually considered for specific IT applications (for example, spreadsheet software, database-management systems, MRP I and MRP II systems, and expert systems), as illustrated in Table 13, but can be generalized

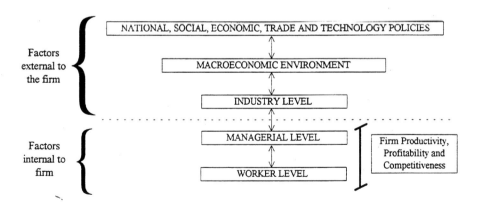

Figure 11. Levels of determinants of technology adoption.

Table 13. Some examples of empirical studies of the adoption–diffusion IT applications.

Authors	Adoption–diffusion phenomenon	Source of data	Adoption–diffusion factors	Major results
Ball et al. (1987): *Data Base*	Adoption of *database-management systems* by industrial firms	Questionnaires from 24 members of the Boston Chapter of the Society for Information Management	*Organizational characteristics* (communication effectiveness, number of engineers and scientists in management, etc.)	Organizations with high R&D commitments and a large number of engineers and scientists in management are more likely to be early adopters
			IT group characteristics (stage in Nolan's life cycle)	
			Information sources (journals, advertisements, salespersons, technical staff, etc.)	
Leonard-Barton (1987): *Interfaces*	Adoption of *SSA* by individual system developers	Survey of 145 programmers, analysts, and supervisors in three sites in a natural-resource firm	*Perceived innovation characteristics* (value, feasibility of use)	Client preferences, adopter attitudes, training in SSA strongly discriminate adopters from nonadopters
			Organizational influences (reward systems, support systems, client preferences)	
			Personal characteristics (demography, skills, years of experience)	Years of experience, perceived accessibility of consulting, supervisor desires, and acquaintance with an advocate are moderately discriminating
Raho et al. (1987): *MIS Quarterly*	Diffusion of *PCs* in industrial firms	Questionnaires from 412 (randomly selected) DPMA members	*Educational commitment* (uncommitted, passive, active, strategic as per McFarlan and McKenny's model)	Phase of diffusion significantly related to level of educational activities

Study	Description	Method	Variables	Findings
Leonard-Barton and Deschamps (1988): *Management Science*	Adoption of an *expert system* by individual sales personnel	Telephone survey of 93 salespeople in dozens of sales sites of a multinational computer company	*Personal characteristics* (innovativeness, job-determined importance, subjective importance of task, task-related skills, software-use skills, sales performance) *Managerial influences* (perceived management support, management urging)	Management was more likely to be viewed as having "suggested" or "required" use of the system by people rating "low" on all personal characteristics (except software-use skills)
Davis (1989): *MIS Quarterly*	Study 1: Current use of *mainframe productivity software* by white-collar workers Study 2: Predicted future use of *PC graphics software* by MBA students	Study 1: Questionnaires from 112 users in IBM Canada's development laboratory Study 2: Questionnaires from 40 students attending a large university	Studies 1 and 2: *Perceived technological characteristics* (perceived usefulness, perceived ease of use)	Study 1: Perceived usefulness and ease of use, both highly correlated with self-reported current use Study 2: Perceived usefulness and ease of use, both highly correlated with self-reported predicted future use In both studies, ease of use appears to be a causal antecedent of usefulness, with little direct effect on use
Davis et al. (1989): *Management Science*	Current use and predicted future use of a *word-processing package* by MBA students	Two waves of questionnaires (14 weeks apart) from 107 MBA students attending a large Midwestern university	*Perceived technological characteristics* (perceived usefulness, perceived ease of use) *Expectations of salient referents* *Attitudes* *Behavioural intentions*	Perceived usefulness and ease of use have a significant direct effect on behavioural intentions over and above their effect transmitted through the mediating attitude construct Behavioural intention to use is significantly related to actual self-reported use

(continued)

Table 13 continued.

Authors	Adoption–diffusion phenomenon	Source of data	Adoption–diffusion factors	Major results
Gatignon and Robertson (1989): *Journal of Marketing*	Adoption of *laptop computers* by sales organizations	Questionnaires from 125 senior sales officers in US firms	*Adopter industry competitive environment* (concentration, price intensity, demand uncertainty, communication openness)	Adoption is associated with high vertical integration and high supplier incentives in the supply industry and high industry concentration and low competitive price intensity in the adopter industry
			Supply-side factors (vertical coordination, supplier incentives)	
			Decision-maker characteristics (information preferences and exposure)	Decision-maker characteristics (preference for negative information and exposure to personal information sources) predict adoption
			Organizational characteristics (centralization, selling-task complexity)	
Huff and Munro (1989): *Journal of Information Systems Management*	Adoption of *microcomputers* by individuals	Personal interviews with several dozen microcomputer users	*Perceived innovation characteristics* (relative advantage, compatibility, complexity, triability, observability)	Anecdotal confirmation that microcomputers diffused quickly because of favourable perceived characteristics

Study	Focus	Data	Variables	Findings
Brancheau and Wetherbe (1990): *Information Systems Research*	Adoption of *spreadsheet software* by individual accountants and managers	Questionnaires from 70 accountants and managers in 18 Fortune 1000 firms	*Adopter characteristics* (age, education, exposure to media, interpersonal-communication exposure, opinion leadership, external social participation, etc.) / *Communication-channel types* (mass media or interpersonal) / *Communication-channel sources* (external or internal to company)	Cumulative adoption follows S-shaped curve using logistic function / Early adopters are different from later adopters, as predicted by Rogers (1983) / Mass-media channel types—external sources are more important at the knowledge stage; interpersonal channel types—internal sources are more important during persuasion
Cooper and Zmud (1990): *Management Science*	Adoption and diffusion of *MRP systems* within industrial firms	Telephone survey of 52 members of the American Production and Inventory Control Society	*Innovative characteristics* (task–technology compatibility, technical complexity)	High task–technology compatibility (continuous manufacturing methods, make-to-stock marketing strategies) and low technological complexity (e.g., fewer parts per bill of material and per finished good) positively related to MRP adoption but not diffusion
Gurbaxani (1990): *Communications of the ACM*	Cumulative adoption of the *BITNET computing network* by universities	Quarterly BITNET Network Information Center records and other sources (1981–88)	Adoption modeled as a function of the number of previous adopters and the time	Three functions were used: Gompertz, logistic, and exponential. The logistic clearly provided the best fit with significant t statistics for all model parameters

(continued)

Table 13 concluded.

Authors	Adoption–diffusion phenomenon	Source of data	Adoption–diffusion factors	Major results
Gurbaxani and Mendelson (1990): *Information Systems Research*	Cumulative adoption of *IT* by US firms	Archival data on total IT spending by large US firms from industry publications (1960–87)	Adoption modeled as a function of the level of previous IT spending and the time	Three price-modified functions were used: Gompertz, logistic, and exponential. Confirmed that exponential (price) terms were significant in all three cases, implying that a purely behavioural explanation for IT adoption is incomplete
Kwon (1990): *ICIS Proceedings*	Diffusion of *IT* in the administrative offices of a southeastern university	Field survey of department heads, "opinion leaders," and "MIS co-ordination" for 74 administrative offices	*MIS maturity* (age, applications, equipment) *MIS climate* (management support, user involvement, management attitude) *Work-unit size* *Network behaviour* (centrality, sources, intensity, link sources, link intensities)	External-communication intensity positively correlated with IT diffusion for work groups with a favourable MIS climate
Nilakanta and Scamell (1990): *Management Science*	Initiation, adoption, implementation of *database-requirements analysis and logical-design tools* by industrial firms	Questionnaires from more than 70 lead database designers in 17 Houston-area organizations	Characteristics (perceived utility, skills to use, etc.) of 15 *information sources* (books, periodicals, etc.) and 13 *communication channels* (telephone, library, etc.)	Hypotheses linking characteristics of information sources and communication channels to diffusion not supported (only 12 of 90 regression coefficients significant at P values ranging from 0.05 to 0.15)

Source: Adapted from Fitchman (1992).
Note: DPMA, Data Processing Management Association; IBM, International Business Machines Corp.; ICIS, International Conference on Information Systems; MBA, master of business administration; MIS, management information science; MRP, material-requirements planning; PC, personal computer; R&D, research and development; SSA, structured systems analysis.

to other IT applications. Some factors listed in Table 13, such as individual characteristics, attitudes, or influences, also pertain to the decision-making process, which will be examined in Chapter 3.

As outlined in Table 13, each empirical study shed light on a particular set of adoption factors in some specific contextual environment. The following two sections will attempt to present an exhaustive list of factors. Please note that a particular factor will be mentioned only if it played a significant role in more than one empirical study.

Internal factors (at the firm level)

Factors internal to the firm that may affect the adoption of technologies can be grouped into three categories: the firm's past experience with technology, the firm's characteristics, and the firm's pursued strategy.

Firm's past experience with technology

A firm's past experience with technology in terms of exposure and organizational learning ultimately affects its future choices in adopting technology (Burgelman and Rosenbloom 1989). This past experience can be captured through notions such as time since first acquisition, number and type of technologies or applications adopted, percentage of different classes of personnel familiar with the technologies, and the current level of assimilation and integration of the technologies.

Firm's characteristics

A firm's characteristics include size, but the influence of this on the adoption of IT applications remains unclear. The adoption of certain technologies may appear more appropriate for larger firms because of the generally large capital investments required and the skilled human resources involved in the implementation and operation of such technologies, but a strong case can be made for successful adoption by smaller firms. They react more quickly, both internally and externally, than larger firms because they are less affected by organizational inertia and because they show a greater degree of involvement by organizational members — especially top management — during implementation. Finally, readily available software and less expensive equipment are now making IT applications more attractive to smaller firms. However, the availability of financial resources (which is associated with size) can be a major stumbling block to the adoption of sophisticated technologies.

Table 14. Summarizing the different dimensions of the internal factors.

I: Firm's past experience	II: Firm's characteristics	III: Firm's pursued strategy
Time since first acquisition	Availability of financial resources	Strategic orientation: Aggressive
Number of technologies or applications adopted	Centralization	Analytic Defensive Futuristic
Types of technologies or applications adopted	Formalization	Proactive Risky
	Technocratization	
Current level of assimilation and integration of technologies	Size	Technological policy
		Technological awareness
Percentage by class of personnel familiar with the technologies		Technological scanning

Structural characteristics can be captured by indicators such as the degree of centralization in the firm, the degree of formalization of the different activities in the firm, and the degree of technocratization, which measures the percentage of technical employees in the firm. All these characteristics[6] have been shown to be associated with the adoption of technology, particularly technocratization, which is a strong contributing factor.

Firm's pursued strategy

Our third group of internal factors deals with the firm's pursued strategy in both strategic orientation and technological policy. A firm's strategy reflects its actions vis-à-vis markets and technology, which ultimately modify its experience and consequently its overall characteristics and capabilities. The need for a strong technology–strategy connection (or fit) has been advocated by a number of authors (see, for example, Powell 1992), and investments in IT should therefore be closely aligned with overall corporate strategy.

These internal factors are summarized in Table 14. It should be noted that all of the perceptual measures found in the second and third groupings in Table 14 use multi-item constructs; these will be presented in "Operational Measures for Internal and External Adoption Factors."

---

[6]   A case for the reverse can be made here as well: the use of IT also has a definite impact on the organization of firms (Rabeau et al. 1994). The feedback loop provides an explanation for this phenomenon.

External factors

External factors are conditions that exist in a firm's external environment and may affect its technology-adoption decisions. These factors can be found at the industry level, in the macroeconomic environment, or in national policies.

Industry level

At the industry level, we are looking at characteristics such as the degree of diffusion of certain technologies, the availability of external know-how (for example, technology suppliers), the degree of innovativeness of the industry, the requirements imposed by major customers and external markets, and overall levels of competition and technological sophistication in the industry.

Macroeconomic environment

Regarding the macroeconomic environment, the concern is more with the availability of certain conditions such as capital and qualified human resources, as well as issues related to the general characteristics of the work force and the type and quality of industrial relations.

National policies

When considering national policies, we must look for actions that may ultimately affect technology adoption in a nation. These actions may come as a result of national policies implemented by the host nations — for example, tax policies, such as investment tax credits aimed at making adoption easier or more accessible to certain groups of firms; or trade agreements between nations, such as the North American Free Trade Agreement (NAFTA), which modify the competitive environment and force firms to react to new market conditions. The actions may also be the result of social programs that favour technical education in schools, colleges, and universities. In some countries, such as the United States, defence procurement practices have a significant impact on the technology-adoption practices of the firms that want to do business with government agencies such as the US Department of Defense. Finally, societal values (which can be partially altered by national policies) and cultural effects have a definite influence on the adoption of IT applications. This remains an underinvestigated field of research, and most efforts to date have been limited to the study of cultural differences between Asiatic and Western, English-speaking cultures (for example, Straub 1994). Societal values and culture remain diffuse concepts, and we propose to control for these concepts by simply taking into account the country in which adoption is studied.

Table 15. Summarizing the different dimensions of the external factors.

I: Industry characteristics	II: Macroeconomic environment	III: National policies

Overall competition
  Type of competitors
  Number of competitors
  Proximity of competitors
	Availability of capital
	
	Availability of qualified human resources
		Trade policies (free trade)
		
		Industry regulation

Characteristics of demand
  Type of customers
  Number of customers
  Location of customers
  Sophistication of demand
  Requirements imposed by major
    customers
	Quality of industrial relations
	
	Inflation
	
	Business cycle
		Government buying practices
		
		Defence procurement practices
		
		Technology-adoption tax credits

Degree of diffusion of technologies
  By technology
  By type of competitor
		Corporate taxation
		
		Human-resource training policies and programs

Availability of external know-how from
  Government agencies
  Institutes
  Technology suppliers—vendors
  Trade associations		

Bearing in mind our three levels of intervention, we summarized these considerations in Table 15.

Adoption factors in the specific context of SMEs

Are adoption factors in SMEs similar to the ones that affect larger firms? In general, this seems to be the case, but the relative importance and emphasis assigned to these factors differ. For example, SMEs depend more heavily on external technological know-how. Table 16 offers a summary of some empirical studies of adoption–diffusion of IT applications carried out in SMEs.

Operational measures for internal and external adoption factors

This section will present ways to measure the importance in a firm of the different factors listed in Tables 14 and 15. Some factors are factual (for example, time since first acquisition of an IT application), whereas others are perceptual (for instance, technological awareness). All perceptual factors are measured using multi-item constructs that have been previously tested, and exact references for these constructs are given in Table 17. However, references for factual factors are not offered because they represent traditional measures.

Table 16. Some examples of empirical studies of the adoption–diffusion of IT applications in the specific context of SMEs.

Authors	Adoption–diffusion phenomenon	Source of data	Adoption–diffusion factors	Major results
DeLone (1988): *MIS Quarterly*	Successful use of CBIS	Two question-naires sent to two respondents (one for the CEO and the other for the person responsi-ble for information systems) (*n* = 93 small manufac-turing firms)	CBIS success related to greater use of exter-nal programing support, higher levels of CBIS planning	
			CEO with greater computer knowledge	CEO is key to the realization of potential impact
			CEO who is more deeply involved	
			Higher levels of computer acceptance by employees	
			More sophisticated computer controls	
			Greater computer training, on-site computers	
Raymond (1985): *MIS Quarterly*	Computerized firms	Questionnaire addressed to the controller in small manufacturing firms (*n* = 464)	Relationship between organizational character-istics (EDP experience, development, opera-tion, application, interface, MIS rank) and user satisfaction and system use	Firm size is not associated with user satisfaction or system use
				MIS success is related to sophistication level of applications
Alpar and Ein-Dor (1991): *Information and Management*	Information-system concerns	Questionnaire mailed to small high-technology firms (*n* = 636)	Ranking information-system concerns (reliability, system quality, change, cost, development, integration, control, people, data management, hardware, software)	Major concerns about information systems are system reliability, system quality, change, cost, development, integration, and control

(continued)

Table 16 concluded.

Authors	Adoption–diffusion phenomenon	Source of data	Adoption–diffusion factors	Major results
Lefebvre and Lefebvre (1992): *Journal of Engineering and Technology Management*	Adoption of 28 computer-based administrative and manufacturing applications	Self-administered questionnaire from CEOs of 95 small manufacturing firms	CEO's personal characteristics	CEO characteristics and degree of firm innovativeness are closely related
			CEO's attitudes and personality traits	Variables related to the CEO are more important than structural characteristics
			Characteristics of CEO's decision-making process	
			Firm's characteristics	
Doukidis et al. (1993): *International Conference on Information Systems*	Use of microcomputers for at least 6 months and up to 5 years	Semistructured interviews in 50 small firms in Greece (retail and distribution, services and manufacturing); longitudinal study (1984 vs 1989)	Previous experience with computers	Improved processing and availability of information and time savings are the most important adoption factors
			Factors influencing the initial decision to computerize (improved availability of information, time savings, improved stock-control procedures, improved accounting procedures, cost reduction, staff reduction)	
			Major obstacles to adoption (lack of computer experience, software and hardware selection, implementation problems, cost and adequate service)	Lack of computer experience is the major stumbling block

Source	Focus	Method	Factors	Findings
Kirby and Turner (1993): *International Journal of Retail and Distribution Management*	Use of computers in 147 small retail businesses	Survey questionnaire	Computer literacy of CEO Dependence on supplier Failure to appreciate the hard and soft benefits of IT	The greater the computer literacy of a small-business owner and the greater the dependence on the supplier, the more likely the firm will adopt IT through awareness of its benefits, especially for the strategic management of the business
Palvia et al. (1994): *Information and Management*	Adoption of computers, use of software packages and information systems	Questionnaires mailed to very small firms with fewer than 50 employees ($n = 131$)	Business characteristics (size, age, profitability) Individual characteristics of CEO (general education, computer knowledge)	Strongest determinants of computing in very small firms are size, computing skills of the owner–manager, and age of the business
Lefebvre et al. (1996): *IEEE Transactions on Engineering Management*	Adoption of advanced computer-based production applications (actual use and planned adoption)	Questionnaires mailed to SMEs ($n = 116$ independent SMEs)	Technical capabilities (of different categories of employee) Strategic motivations (in terms of costs, productivity, quality, flexibility) Influences of internal and external proponents	Strongest determinants of future adoption of advanced manufacturing technologies are the technical capabilities of blue-collar workers, the influence of customer and technology suppliers, and outward-oriented strategic motivations

Note: CBIS, computer-based information system; CEO, chief executive officer; EDP, electronic data processing; MIS, management information science; SMEs, small and medium-sized enterprises.

Table 17. Operational measures for internal and external adoption factors.

Internal factors (Table 14)	Measures
Firm's past experience	
Time since first acquisition	Exact date
Number of IT applications adopted	Simple count of IT applications, as listed in Table 12
Current level of assimilation and integration of technologies	Score, as proposed in "Proposed Measurement of the Level of Adoption of IT Applications"
Percentage by class of personnel familiar with the technologies	Percentage of employees actually using IT by category (clerical staff, secretarial staff, managers, professionals, blue-collar workers)
Firm's characteristics	
Size	Volume of annual sales (actual and projected for each of the next 3 years)
Availability of financial resources	Actual amount spent on IT applications (hardware and software) on an annual basis
	Budgeted amount on IT applications for each of the next 3 years
	Relative importance of investments in IT applications (actual amount divided by annual sales; budgeted amount divided by projected annual sales)
Centralization	Multi-item construct first proposed by Miller and Friesen (1982) (see Appendix C)
Formalization	Multi-item construct used by Lefebvre and Lefebvre (1992) (see Appendix C)
Technocratization	Number of scientists, engineers, programmers, and technicians divided by the total number of employees
Firm's pursued strategy	

Strategic orientation
    • Aggressiveness
    • Analysis
    • Defensiveness
    • Futurity
    • Proactiveness
    • Riskiness
	The six dimensions of the strategic orientation are captured by multi-item constructs proposed by Venkatraman (1989) (see Appendix C)

Technology policy	Multi-item construct proposed by Ettlie and Bridges (1987) and adapted to the context of SMEs in Lefebvre et al. (1993) (see Appendix C)
Planned introduction of IT applications	Planned introduction of the applications listed in Table 12 within the next 3 years
Technological awareness	Multi-item construct first used by Miller and Friesen (1982) (see Appendix C)

(continued)

Table 17 concluded.

External factors (Table 15)	Measures
Industry characteristics	
Overall competition	
Type of competitors	Direct competitors: mainly small, medium-sized, or large firms or multinationals
Number of competitors	Number of direct competitors
Proximity of competitors	Number of direct competitors in the region, in the country, and outside the country
Characteristics of demand	
Type of customers	Industrial vs individual customers; mass vs customized products or services
Number of customers	Average number of customers by type
Location of customers	Average number of customers in local, national, and international markets
	Percentage of sales realized in local, national, and international markets
Sophistication of demand	Perceived level of sophistication of different types of customers
Requirements imposed by major customers	Perceived ease of predicting customers' demand and requirements
Degree of diffusion of technologies • By technology • By type of competitor	Comparison with existing national statistics (if available)
Availability of external know-how	Accessibility, usefulness, and cost of external know-how from agencies, institutes, technology suppliers–vendors, and trade associations
Macroeconomic environment	
• Availability of capital (including venture capital) • Availability of qualified human resources • Quality of industrial relations • Inflation • Business cycle	Porter (1980) discussed a wide variety of competitive advantages different countries can create for their firms; these measures are country specific
National policies	
• Trade policies (free trade) • Industry regulation • Government buying practices • Technology-adoption tax credits • Corporate taxation • Social and economic policies • Human-resource training policies and programs	Once again, measures are country specific. Baldwin et al. (1994) proposed a list of governmental programs and actions that were rated by firms (in terms of relative importance)

Note: SMEs, small and medium-sized enterprises.

Internal factors are universal, or generic, and their measurement can therefore be useful in any country. However, external factors (especially the ones pertaining to the macroeconomic environment and to national policies) are country specific. For international comparisons, identifying a common set of measures for different countries poses certain difficulties.

CHAPTER 3

CHARACTERISTICS OF THE DECISION-MAKING PROCESS AS A PRIME ADOPTION FACTOR

The very way in which the decision to adopt ITs is made can be one of the strongest determinants of the actual adoption of these technologies (Dean 1987; Pennings 1987; Lefebvre and Lefebvre 1992).

Rogers (1983) proposed one of the best-known models of the innovation-adoption process. According to Rogers, the process can be divided into five stages: (1) becoming aware of the innovation under consideration; (2) forming a favourable or unfavourable attitude toward it; (3) deciding to adopt; (4) implementing the innovation; and (5) deciding whether to keep the innovation after it has been implemented.

In considering the context of the adoption of ITs, we will be especially interested in the first three stages. During the first stage (that is, the "intelligence," or research, stage) systematic scanning of information is of the utmost importance, and the availability of technical information from different sources, as well as from knowledgeable or influential proponents (internal or external), therefore plays a major role. During the second and third stages, or the evaluation and choice stages, the perceived characteristics of the IT application being considered, as well as the credibility of the internal and external proponents, may strongly influence the outcome of the decision-making process. Although the decision-making process is less articulated and less formal in SMEs than in larger firms, it is present even in the very smallest firms.[7]

Based on the above discussion, we propose that the process underlying technology-adoption decisions be captured along five major dimensions: the influence of internal proponents; the influence of external proponents; the availability

---

[7] For example, small firms rarely perform a cost–benefit analysis before adopting an IT application, and the evaluation is based on expected outcomes (which are sometimes grossly underestimated or overestimated). Small firms also tend to rely very heavily on technology vendor–suppliers (Gerwin and Kolodny 1992; Lefebvre et al. 1996).

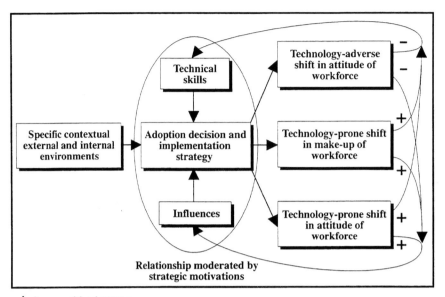

+ has a positive impact on
− has a negative impact on

Figure 12. Evolving influences of the work force on IT applications. Source: Lefebvre et al. (1996).

of technical information; the perceived characteristics of an IT application; and the justification process for the adoption of an IT application. These five dimensions will be treated separately, although in reality they can rarely be dissociated.

Influence of internal proponents

If one characterized the main driving force in SMEs, it would have to be the people who make these productive units work and succeed with them. One important player is the CEO, who not only runs the firm but also provides direction in beliefs, vision, and day-to-day operating procedures. The CEO nevertheless needs support from competent people to achieve the desired goals. These employees obviously have inherent characteristics, such as age, experience, and know-how, but they also have less measurable characteristics, such as motivation or the ability to become involved in change. The underlying dynamic of the adoption of technology is complex, as illustrated in Figure 12, and it evolves.

The first wave of technological change may meet with stiff resistance because fear of the unknown (even deskilling and job destruction) is a common reaction. As the reality of the adoption of IT applications sinks in, however, employees whose skills (because of their age, attitudes, or aptitudes) cannot be upgraded leave; others are trained and become more and more knowledgeable

(Adler 1986; Zammuto and O'Connor 1992); and other qualified workers are gradually added to the work force. Thus, the work force as a whole becomes more receptive to technological change. Indeed, the work force may gradually be transformed from an obstacle to technological change (technological retardant) into a technology-neutral factor, even into a factor inducing further adoption of IT applications.

The influences of all internal proponents (or the work force) will evolve as IT adoption becomes more and more intense (Figure 12), but these people will also play different roles, depending on their functional expertise (Dean 1987). For instance, production and marketing managers will emphasize different potential benefits when considering the adoption of IT technologies. Furthermore, all these internal proponents not only will be involved in the actual decision to go ahead with adoption but also will be greatly responsible for the success of the implementation. For these reasons, one must take a careful look at the different characteristics of all the workers, along with the CEO, because some may promote and others may impede the adoption of technology in the firm. These characteristics can be listed as follows:

CEO characteristics
- Age;
- Level of education;
- Type of education;
- Functional experience;
- Computer literacy and experience;
- Attitude toward risk;
- Influence on initiating the idea of adopting; and
- Influence on the actual decision to adopt.

Employees' characteristics
- Age of workers by category;
- Employee attitude and motivation toward technology by category;
- Influence of the different categories of worker on initiating the idea of adopting; and
- Influence of the different categories of worker on the decision to adopt.

Influence of external proponents

Firms are continuously subjected to external pressures. This is also the case where the adoption of technology is concerned. The literature has extensively documented the fact that SMEs must often engage in the adoption of technology because their major customers require them to do so (Hill 1994). In certain industrial sectors, large contractors will force their suppliers to operate under specific guidelines using various forms of technological support, such as EDI, and this accelerates the rate of vertical integration in a specific sector of economic activity.

Similarly, pressures from competitors and even from trade associations will oblige firms to acquire new technology to maintain their competitive position. The adoption of IT applications will also be moulded to a large extent by vendors of hardware and software (especially in SMEs; see, for example, Baker 1987) and by consultants because these two types of external proponents will emphasize certain positive characteristics of IT applications. Among these characteristics, perceived usefulness and ease of use (Davis 1989), perceived relative advantage, compatibility, complexity, triability, and observability (Huff and Munro 1989) are the focus of technology vendors and consultants. Finally, governments may be influential by aggressively promoting IT adoption or creating incentives for such adoption.

In sum, the external proponents can be listed as follows:

- Customers and contractors;
- Direct competitors;
- Trade associations;
- Hardware and software vendors;
- Consultants; and
- Governments.

The relative importance of these external proponents will vary not only from sector to sector, but also sometimes from country to country.

Availability of technical information

One important aspect of IT adoption is the availability of information concerning specific technologies. This information is essential when developing diffusion policies. The following is a list of the sources of technical information, derived from our literature review:

External sources
- Trade shows;
- Professional journals;
- Technology suppliers and vendors;
- Consultants;

- Customers;
- Bankers;
- Competitors;
- Industrial partners;
- Government agencies;
- Trade associations; and
- Colleges, technical schools, and universities.

Internal sources
- R&D group (if it exists);
- Marketing group;
- Production group;
- Accounting–finance group; and
- IT group (if it exists).

External sources of information were the most widely cited, but internal sources may also prove useful. Indeed, internal proponents not only are influential during the decision process (see "Influence of Internal Proponents") but also are active as prime sources of technological information. The technological gate-keepers, or boundary specialists (usually computer specialists, engineers and employees or CEOs with an ongoing curiosity about new technological developments), can be valuable (and relatively cost free) sources of information.

Irrespective of the source, there is clearly a contextual dimension related to the importance of all these sources that takes into account their existence and accessibility and relative cost of consultation. This should be monitored closely for all the sources of technical information.

Finally, leaders of the most innovative firms tend to assign greater importance to a widespread internal and external search for technological information.

Perceived characteristics of an IT application

Characteristics of IT applications explain their rate of adoption and diffusion. On the basis of a very well known and thorough analysis of more than 1 500 studies on the diffusion of innovations, Rogers (1983) emphasized the crucial importance of perceived[8] characteristics or attributes of innovations when predicting the future rate of adoption. Five attributes that were "as mutually exclusive and as

---

[8] The notion of perceptual rather than factual characteristics should be emphasized here. Perceptions, no matter how distorted they may be (Byod et al. 1993), override reality (Lefebvre et al. 1996).

universally relevant as possible" (Rogers 1983, p. 211) were shown to influence the rate of adoption: relative advantage, compatibility, complexity, triability, and observability. These five attributes were shown to influence the adoption of IT applications, which is known as one form of innovation, namely, process innovation (Pennings 1987).

When a specific IT application is perceived as having low compatibility (with existing values, past experiences, needs of potential users, and implemented technologies), high complexity, low triability, and low observability, they are considered radical (see "Characteristics of IT Applications: Primary Versus Secondary Attributes"). Relative advantage (or the degree to which an IT application is perceived as better) can be assessed in terms of its economic–financial and competitive advantages (see "Overemphasis on Financial Feasibility").

Roger's (1983) five characteristics were found to be associated with the adoption of microcomputers (Huff and Munro 1989). Other characteristics were also found to be associated with the adoption of some specific IT applications: for example, perceived usefulness and ease of use (Davis et al. 1989), perceived utility (Nilakanta and Scamell 1990) or task–technology compatibility,[9] and technical complexity (Cooper and Zmud 1990). Finally, cost, which is often underestimated in SMEs because implementation costs are not always accurately estimated, is definitely the major factor hampering the adoption of technology in these firms. However, cost is relative to the context of each firm.[10]

Justification process for the adoption of an IT application

The literature on the justification for advanced computer-based technologies is extremely voluminous. The relevant information is scattered throughout numerous scholarly and professional journals. For instance, Son (1992) published a comprehensive bibliography on the justification for AMTs, listing 274 articles from 92 sources (77 journals and magazines and 15 conference proceedings; books and unpublished working papers were excluded).

After scanning such a vast literature, we can make two main observations: first, there is a need for an interactive top-down–bottom-up decision-making process; second, the importance of financial feasibility overrides that of organizational and technological feasibility.

---

[9] These two attributes cover some aspects of compatibility and complexity, as described by Rogers (1983).

[10] What is considered rather expensive by one firm may be considered affordable by another firm of similar size but in a different competitive position.

Top-down–bottom-up decision-making process

Most small firms are characterized by a top-down decision-making process, that is, from the CEO down through the organization. This process can be beneficial by speeding up decisions that might otherwise be difficult to make, but because of the very nature of some of the intangible benefits associated with the adoption of technology, such a process can also become a barrier to successful adoption and dissemination. Most of the literature on organizational change stresses the importance of employees' participating and buying in when entering into a process of change like the one technology generates in most circumstances. This supports more of a bottom-up approach, whereby employees are involved in the decision to adopt and in the actual implementation process. It would thus appear that a mixture of both top-down and bottom-up decision-making processes would best ensure the successful adoption of technology in small firms.

Overemphasis on financial feasibility

In the evaluation–choice, or justification, stage, many CEOs and top managers appear to place too much emphasis on short-term profitability. This emphasis on short-term profit may hinder the adoption of new technology. Moreover, top managers who do not get involved in the technological process tend to underestimate the financial resources to be allocated to technology-related activities and favour those divisions that are likely to bring in immediate profits.

In addition, other authors have noted that the measures used to evaluate these profits have been inadequate. For example, in a classic study, Dearden (1969) denounced the dysfunctional aspects of the emphasis on return on investment (ROI). Many researchers (in particular, Finnie 1988; Kumar and Loo 1988) maintained that the financial evaluation techniques used during the process of deciding to acquire new manufacturing technologies constituted in themselves major obstacles to the process.

Finally, the basic applications (such as transactional systems) can provide very tangible benefits, usually in the form of expected cost savings. However, the most advanced IT applications may also increase profits and revenues (through gaining competitive advantages, such as improved flexibility or increased quality) or reduce the investments[11] required to sustain daily activities (such as stock control and planning). The main problem resides in correctly assessing these benefits, especially the intangible ones such as the quality of customer services. This consti-

---

[11] For example, JIT can radically reduce inventory levels, and a firm's capital investment in stock will therefore be much lower.

tutes a major problem in SMEs because they may not keep track of the different indicators[12] of competitive advantage. Financial feasibility is strongly biased toward the benefits that can be more easily identified.

Organizational and technological feasibility

Beyond financial feasibility, one must consider organizational feasibility, that is, the capacity of the organization to adopt and successfully implement the new technologies considered. More specifically, does the organizational culture favour such change? Are the employees ready for and open to these changes? Are proposed changes in line with the current activities of the organization? Is it the right time to make them? Will these changes modify the employees' tasks, the organizational structure, the relationship with suppliers and clients? All these questions need to be addressed before the final adoption decision, and when overlooked they can lead to disastrous situations, even when financial feasibility is positive.

Finally, one must evaluate the technological feasibility in terms of compatibility with existing in-house systems and those in the industry, level of complexity, dependability, ease of repair, quality of service, and upgrading possibilities. This evaluation can be done by experts in the firm, technology suppliers, and ultimately, other users of similar technologies. Technological feasibility in SMEs is fairly difficult to assess because of the lack of internal expertise. In fact, the main factors that tend to hinder the adoption of IT applications in small firms were found to be as follows (Baker 1987):

- A feeling that technology was changing too rapidly;
- Lack of confidence in claims made by computer sales representatives;
- Lack of time to do the necessary analysis to determine what or how to automate;
- Lack of confidence in the ability of computer vendors to provide ongoing service and support after implementation; and
- Lack of knowledge of new technology.

These five factors indicate how difficult it is for SMEs to address the issue of technological feasibility, which is, of course, essential to the successful adoption and implementation of new technologies.

In sum, the following are the four main characteristics of the justification process:

---

[12] Basic indicators, such as average number of rejects or average delivery time, are not monitored in some SMEs.

- Top-down–bottom-up decision-making process;
- Financial feasibility;
- Organizational feasibility; and
- Technological feasibility.

A top-down decision-making process and too much emphasis on financial feasibility may well hamper IT adoption, whereas a bottom-up decision-making process and organizational and technological feasibility are closely related to the successful adoption of IT applications. However, if financial feasibility is overlooked, small firms face many difficulties, especially during the implementation phase.

Operational measures for characteristics of the decision-making process as a prime adoption factor

Each characteristic of the decision-making process has to be measured. Table 18 (pp. 54–56) summarizes the operational measures for each of these characteristics.

One may wish to analyze in greater detail the relative importance of the different sources of technical information. It is possible to assess the accessibility, usefulness, credibility, and cost of each of these sources (although cost is difficult to estimate for internal sources of information). However, this might be a rather lengthy exercise because 17 sources are considered here. In general, the characteristics of the decision-making process are better captured by on-site interviews with multiple respondents (including the CEO) because of the behavioural aspect and the complexity of these characteristics.

Table 18. Operational measures for the characteristics of the decision-making process with respect to IT adoption.

Influences of internal proponents	Measures
CEO's characteristics	
Age	Date of birth
Level of education	High-school diploma College diploma University studies (certificate) BA, MA, or PhD
Type of education	Technical or scientific Nontechnical or nonscientific
Functional experience	Length of experience in the firm (as CEO and as employee) and in the sector
	Length of experience in accounting–finance, sales–marketing, engineering–production, and human-resource management
Computer literacy and experience	Length of time since first introduction to computers
	Frequency of use of IT applications
Attitude toward risk	Multi-item construct (see Appendix D)
Influence on initiating the idea of adopting	Relative importance of such influence
Influence on actual decision to adopt	Relative importance of such influence
Employees' characteristics for each category (clerical employees, secretarial employees, managers, professionals, and blue-collar workers)	
Age	Average age of the work force
Level of education	Percentage of employees with high-school diploma, college diploma, etc.
Computer experience and literacy	Percentage of employees with previous experience with computers
Attitude and motivation toward technology	Positive vs negative attitude
	Level of resistance to change
	Motivation to learn and use IT applications
Influence on initiating the idea of adopting	Relative importance of such influence
Influence on the actual decision to adopt	Relative importance of such influence

(continued)

Table 18 continued.

Influences of external proponents	Measures
Influence on initiating the idea of adopting	Relative importance of such influence of each of the following groups: Customers and contractors Direct competitors Trade associations Hardware and software vendors Consultants Governments
Influence on actual decision to adopt	Relative importance of such influence of each of the following groups: Customers and contractors Direct competitors Trade associations Hardware and software vendors Consultants Governments

Availability of technical-information sources	Measures
Availability of external sources	Relative importance of each of the following sources of technical information: Trade shows Professional journals Technology suppliers and vendors Consultants Customers Bankers Competitors Industrial partners Government agencies Trade associations Colleges and technical schools Universities
Availability of internal sources	Relative importance of each of the following sources of technical information: R&D group (if it exists) Marketing group Production group Accounting–finance group IT group (if it exists)

Characteristics of IT applications	Measures
Attributes of IT applications being considered for adoption	Perceived characteristics of IT applications, as proposed by Rogers (1983) Relative cost

(continued)

Table 18 concluded.

Characteristics of the justification process	Measures
Top-down–bottom-up decision-making process	Relative importance of internal proponents (see Appendix D)

Financial feasibility	Existence of a written (formal) financial feasibility plan:
	    Cost–benefit analysis
	    ROI
	    Net present value
	    Assessment of intangible benefits

Organizational feasibility	Existence of a written (formal) organizational feasibility plan
Technological feasibility	Existence of a written (formal) technological feasibility plan

Note: CEO, chief executive officer; R&D, research and development; ROI, return on investment.

CHAPTER 4

IMPACTS OF ADOPTION

There is no doubt that the impact of IT is significant. However, "the real challenge is not technology (adoption) per se, but the ability to adapt to take advantage of its emerging functionality" (McKenny 1995, p. 37). Reaping the full benefits of IT adoption requires not only a full understanding of IT applications and their potential but also a readiness to change, all of which points to the importance of mobilizing human resources and constantly improving technical capabilities.

Moreover, as more and more firms successfully adopt and implement IT applications, the comparative competitive advantages derived from the adoption of these applications may very well disappear if firms do not stay ahead. For firms that lag behind, IT adoption becomes merely a question of survival.

This partially explains why contradictory results concerning the impacts of IT adoption have been observed. As we will explain, IT adoption as such is a necessary but not sufficient condition for increased productivity, key competitive and strategic benefits, and stronger financial and export performance. We will also try to assess the ambivalent effect of IT on work and employment.

Relationship between IT and productivity: the "elusive connection"

The IT–productivity paradox has generated considerable interest among practitioners, professionals, and theorists, especially economists. Because productivity is the fundamental economic measure of IT's contribution, it should be closely examined at the macrolevel and the industry level, as well as at the microlevel (or firm level).

Relationship between IT and productivity at the macrolevel and the industry level

In the last 10 years or so, capital spending was steady for industrial equipment, whereas it constantly increased for computers and communication equipment in the industrialized nations (see Figure 13 for the United States).

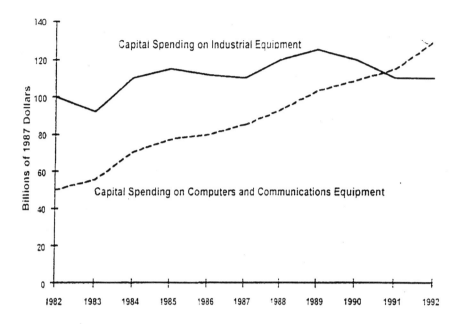

Figure 13. Rise of capital spending on ITs in the United States (1987 US dollars). Source: *Fortune* (13 Dec 1993).

As a result, computing power rose sharply in many countries, with a striking increase registered in the United States (Figure 14).

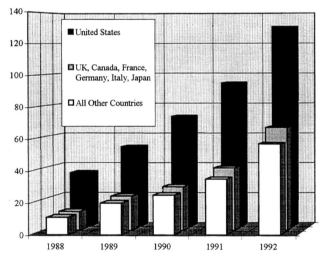

Figure 14. Worldwide growth in computing power. Source: *Fortune* (13 Dec 1993).

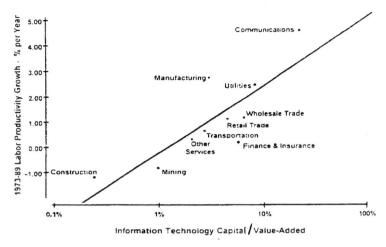

Figure 15. National productivity indicators based on capital assets in the United States. Source: NRC (1994).

Despite these large investments, IT did not seem to help national productivity, even in the United States, where these investments were the most evident. As Robert Solow bluntly put it, "we see computers everywhere but in the economic statistics." [13]

If, instead of looking at productivity at the national level, we turn to specific sectors, we find a very different profile, depending on the specific sector (Figure 15). Communications, manufacturing, and utilities demonstrated the largest increases in labour productivity as measured by the ratio of IT investments to the value added for that sector. However, IT investments (ranging from 1 to 10% of the value added) in wholesale and retail trade, transportation, finance and insurance, and other services led only to marginal increases in labour productivity (between 0 and 1%). Mining and construction lagged behind, having the lowest rates in IT investments, along with negative productivity growth.

Thus, it appears that the IT–productivity paradox is centred not on manufacturing but on the services sector. Overall, the US services sector spent more than 862 billion United States dollars (USD) in the last 10 years (representing about 85% of total US IT-hardware investments), but its average productivity growth was only 0.7%, significantly lower than the rate for the manufacturing sector. Furthermore, for the federal agencies, which also invested massively in computer hardware, software, networks, and peripherals in the last decade, total government output per employee–year from 1982 to 1990 only grew

---

[13] Quote from Nobel laureate economist, Robert Solow, from the *MIS Quarterly* (Jun 1994): editor's comments.

at a rate of 0.8%,[14] according to the US Bureau of Labor Statistics. What was happening? This paradox can be explained by a number of factors:

1.   The measures used in conventional approaches are faulty. In particular, the traditional measurement of productivity, based on the ratio of output to input units, is arbitrarily set and not necessarily comparable from one type of service (for example, health services) to another (for example, financial services). The economy-wide productivity data produced by the US Bureau of Labor Statistics are also subject to severe criticism because 58% of service industries are not included in the national figures and because measurements are highly suspect in other service industries. Finally, when the standard measurement of output is expressed by profit or sales, highly deregulated industries, such as transportation, may show low productivity because their profit margins have shrunk, although their customer services have drastically improved. Multifactor or total-factor productivity may eliminate some biases and distortions by taking into account several inputs (labour, capital, and materials) for a given output, but the lack of data severely limits the availability of national indicators.

2   The distinction between the manufacturing sector and the services sector may be misleading because many manufacturers outsource their less productive activities, which are often related to services.

3.   A lag effect may occur because IT adoption and implementation require extensive learning and adjustment. Benefits from IT may take several years to become apparent (Brynjolfsson 1993) and are cumulative (Lefebvre et al. 1995).

4.   Mismanagement of IT is another possible explanation. The implementation of IT must not simply superimpose new technology on old processes. IT is also particularly vulnerable to misallocation and overconsumption by managers.

5.   Redistribution and dissipation of benefits may occur at the macrolevel. IT adoption may be beneficial to individual firms but may be unproductive

---

[14] For the longer period from 1967 to 1990, the rate was 1.4%. Therefore, investments in IT did not seem to be accelerating productivity growth; rather, they seemed to be having the opposite effect.

Table 19. Principal empirical studies of IT and productivity.

Economy wide or cross sector	Manufacturing	Services
Osterman (1986)	Loveman (1988)	Cron and Sobol (1983)
Baily and Chakrabarti (1988)	Weill (1990)	Franke (1987)
Lefebvre and Lefebvre (1988)	Morrison and Berndt (1990)	Harris and Katz (1989)
Roach (1989)	Barua et al. (1991)	Roach (1989)
Brooke (1991)	Siegel and Griliches (1991)	Alpar and Kim (1990)
	Brynjolfsson and Hitt (1993)	Noyelle (1990)
		Parsons et al. (1990)
		Strassmann (1990)
		Brynjolfsson and Hitt (1993)

Source: Adapted from Brynjolfsson (1993).

from the standpoint of the industry as a whole. (The metaphor usually used is "IT rearranges the shares of the pie without making the pie any bigger.")

We have listed the five most frequently mentioned causes of the IT–productivity paradox at the macrolevel. However, we strongly believe that the relationship between IT and productivity would be better observed at the firm level. The choice of the firm as the unit of analysis can be easily justified by the fact that no country is able to raise its national productivity over the long term without doing so at the firm level.[15] Furthermore, any national call for increased productivity will produce results only if concrete actions are undertaken in firms. Let us therefore examine the relationship between IT and productivity at the firm level.

Relationship between IT and productivity at the firm level

One of the first studies of the impact of IT was carried out by Cron and Sobol (1983). In a sample of 138 medical-supply wholesalers, they found that IT was associated with either very high or very low performers. This classic study led to the hypothesis that IT tends to reinforce existing management practices, whether they are efficient or inefficient. Several empirical studies are listed in Table 19.

---

[15] This is, in fact, part of Porter's message when he mentions that "firms, not nations, compete in international markets" (Porter 1990, p. 33).

Table 20. Studies of IT and productivity in manufacturing firms.

Study	Data source	Findings
Lefebvre and Lefebvre (1988)	667 manufacturing firms	The impact of IT on employee productivity was greater for clerical and secretarial personnel, managers, and professionals than for blue-collar workers
Loveman (1988)	PIMS–MPIT	IT investments added nothing to output
Morrison and Berndt (1990)	BEA	IT's marginal benefit was just 0.80 : 1 USD invested
Weill (1990)	Interviews and surveys	Contextual variables affected IT performance
Barua et al. (1991)	PIMS–MPIT	IT improved intermediate outputs, if not necessarily final output
Siegel and Griliches (1991)	Multiple government sources	IT-using industries tended to be more productive; government data were unreliable
Brynjolfsson and Hitt (1993)	IDG; Compustat; BEA	The return on IT investment was more than 50% per year in manufacturing

Source: Adapted from Brynjolfsson (1993).
Note: BEA, Bureau of Economic Analysis; IDG, International Data Group, Inc.; MPIT, Management Productivity and Information Technology Project; PIMS, Profit Impact of Market Strategy database of the Strategic Planning Institute.

Tables 20 and 21 give some details on the data sources and the main findings of the empirical studies listed in Table 19. Some studies were based on government statistics (for instance, Roach 1989; Noyelle 1990; Siegel and Griliches 1991), but most analyzed data collected from individual firms.

Based on the findings presented in Tables 20 and 21, the IT–productivity connection remains elusive, with contradictory results from study to study. It is therefore necessary to examine the impacts of IT in terms not strictly of productivity but of derived competitive advantage. For example, a firm can be very effective and efficient in its operations but not productive. This would be the case if the firm's products (manufactured goods or services) did not sell as well as expected (because of a recession, for instance); and because the firm's output represents the products sold and delivered, the ratio of input to outputs would remain low even if the firm had definite advantages for key competitive dimensions derived from IT applications.

Impact of IT on key competitive dimensions

The description of information systems as a "competitive weapon" became a cliché during the 1980s. Obviously, information systems do provide support for

Table 21. Studies of IT and productivity in services firms.

Study	Data source	Findings
Cron and Sobol (1983)	138 medical-supply whole-salers	Bimodal distribution among high IT investors was either very good or very bad
Lefebvre and Lefebvre (1988)	996 firms	The impact of IT on employee productivity was greatest for managers in wholesale or retail trade services and for secretarial and clerical personnel in other services firms
Harris and Katz (1989)	LOMA insurance data for 40 companies	Weak positive relationship was shown between IT and various performance ratios
Roach (1989)	Principally BLS, BEA	IT capital per information worker was vastly increased, but measured output decreased
Alpar and Kim (1990)	Federal Reserve data	Performance estimates were sensitive to methodology
Noyelle (1990)	US and French industry	Severe measurement problems occurred in services
Parsons et al. (1990)	Internal operating data from two large banks	IT coefficient in translog production function was small and often negative
Strassmann (1990)	Computerworld survey of 38 companies	No correlation was shown between various IT ratios and performance measures
Brynjolfsson and Hitt (1993)	IDG; Compustat; BEA	The return on IT investment was more than 60% per year in services

Source: Adapted from Brynjolfsson (1993).
Note: BEA, Bureau of Economic Analysis; BLS, Bureau of Labor Statistics; IDG, International Data Group, Inc.; LOMA, Life Office Management Association, Inc.

everyday business operations and activities and are also needed in the diverse decision-making processes in a service organization. They also play a major role in achieving and maintaining or improving strategic advantages (Neo 1988). Bakos and Treacy (1986) noted that there were more than 200 published articles that focused mainly on different frameworks proposing categories of competitive advantage derived from the use of information systems (for example, Porter and Miller 1985) and on success stories.

As technology evolved, numerous articles addressed the topic of EDI or interorganizational systems, which greatly improve customer services, reduce administrative costs, provide faster response to market needs, and allow more flexibility in product design, production, and delivery (Davidow and Malone 1992; De Toni et al. 1994). EDI is thus linked to the emergence of the virtual corporation (Davidow and Malone 1992), closer links binding buyers and suppliers, and

Figure 16. IT and the value chain. Source: O'Brien (1993). Note: SIS, strategic information systems.

higher levels of logistical integration in all activities of the production chain. This value-chain approach, first presented by Porter (1985), represents an important conceptual framework. It emphasizes the sequence of activities that add value to the product. These are the primary activities constituting a firm's core activities related to its product portfolio, as well as the activities that support these primary activities (Figure 16).

As can be observed in Figure 16, all these activities need to be supported by IT applications, some of which are present in entities external to the firm. The model allows a firm to identify its core competencies, the competitive advantages over its direct competitors it wishes to pursue, and therefore the efforts it needs to make in certain activities along the value chain. This becomes particularly important when setting technology-adoption priorities.

To assess the extent of competitive advantage derived from the adoption of IT applications, Sethi and King (1994) identified five distinct dimensions: efficiency, functionality, threat, preemptiveness, and synergy (Table 22). These dimensions and their underlying theoretical concepts apply in the highly sophisticated context of very large firms but not necessarily in the context of SMEs.[16]

The potential benefits of AMTs have also received a great deal of attention. Several authors (Buffa 1985; Singhal et al. 1987; Williams and Novak 1990) argued that AMTs can provide multiple and simultaneous competitive advantages for those seeking cost performance, dependability, quality, and flexibility. North American firms with lower AMT adoption rates seem to lag continuously behind European and Japanese companies (Hottenstein and Dean 1992). However, cases of fruitless automation are well documented (for example,

---

[16] These dimensions and their corresponding set of perceptual measures were tested in the largest manufacturing and service companies in the United States (one data source was Corporate 1000), with top information-systems executives acting as respondents.

Table 22. Benefits derived from IT applications in large firms.

Key competitive dimension	Concepts described in maintenance
Efficiency	Use of IT to reduce cost in functional areas (McFarlan 1984)
	Internal and interorganizational efficiency (Bakos and Treacy 1986)
	Comparative efficiency (Bakos 1987)
	Productivity (Synnott 1987)
Functionality	New products and services (Parsons 1983; McFarlan 1984)
	Customer service (Ives and Learmonth 1984)
	Differentiation (Porter 1985)
	Adding value for customers (Clemons and Kimbrough 1986)
	Unique product features (Bakos 1987; Bakos and Treacy 1988)
Threat	Buyer and supplier power (Parsons 1983)
	Customer and supplier switching costs (Bakos 1987)
	Switching costs and search-related costs (Bakos and Treacy 1988)
Preemptiveness	Preemptive strikes (MacMillan 1983; Clemons 1986)
	Positional advantages and timing (Bakos 1987)
	First-mover effects (Clemons and Knez 1987)
Synergy	Integration with company strategy (King et al. 1986; *Information Week* 1987)

Source: Sethi and King (1994).

see Vieller 1989), and firms, especially small ones, appear to have difficulty gaining the full benefits of these technologies (Schroeder et al. 1989). Nevertheless, if "AMT is not a panacea, it does not have to be a disappointment" (Gerwin and Kolodny 1992, p. 16).

According to Naik and Chakravarty (1992), the acquisition and successful implementation of AMT are now indispensable for manufacturing firms, which have to cope with an increasingly competitive environment characterized by low production costs, smaller batch sizes, product-mix complexity, high-quality products, short product life cycles, and short delivery cycles. There is also evidence that AMTs generate a wide range of strategic advantages, even for small firms (Meredith 1987), resulting in stronger competitive positions for these small manufacturers. The adoption of AMTs therefore greatly improves the competitive performance of most manufacturing firms (Naik and Chakravarty 1992)

The use of computer-based technologies in the nonmanufacturing functional areas, referred to here as administrative applications, also profoundly changes and improves the way these manufacturing firms compete (Bradley et al. 1993). For example, time-based competition, which is now becoming a norm, relies heavily on ITs for the automation of common business processes, such as purchases, inventory, billings, and accounts receivable–payable. In manufacturing firms, it is increasingly difficult to dissociate the impacts derived from administrative and nonadministrative IT applications. Even small manufacturing firms have a mixed portfolio of IT applications (that is, administrative and nonadministrative types of application) in a very high proportion (90.5%), and empirical evidence from these firms shows that the two types of IT application have synergetic impacts on key competitive dimensions (Lefebvre et al. 1995).

Impact of IT on performance

The globalization of markets, capital, and human resources is now an inescapable reality, and internationalization has become a central theme for advocates of economic growth. Broad international trade agreements, such as the one signed by Canada, the United States, and Mexico (NAFTA), are providing firms with new opportunities to expand their market bases.

For SMEs, international markets can be very attractive, and numerous studies have reported that small firms are indeed active outside their national boundaries and capable of facing international competition (Bonaccorsi 1992; Samuel et al. 1992; Thurik 1993).

To what extent does IT contribute to export performance? Although there is ample evidence that IT is the main driving force behind the globalization of capital (almost complete deregulation of worldwide financial markets, instantaneous access to capital from most countries, and capital flows in any currency), the contribution of IT to export performance remains underinvestigated, especially in SMEs. In the age of virtual corporations, there is no doubt that export performance and success depend on "firms' ability to gather and integrate massive flows of information and to act intelligently on that information" (Davidow and Malone 1992, p. 59). Obviously, the link between IT and export performance should, intuitively and logically, be a strong one, especially when one considers the increasing diffusion of real-time manufacturing and service operations.

Financial performance is the bottom line for many firms. However, previous studies failed to convincingly demonstrate a link between financial performance and IT (see, for example, Harris and Katz 1989).

Do firms sacrifice short-term financial benefits by investing massively in IT for the long-term benefits it may provide? Is IT simply a prerequisite for continuing to operate in an increasingly competitive context? These questions have yet to be fully answered.

Impact of IT on work and employment

The impact of IT on the workplace and on organizations has been and will continue to be tremendous. The flattening of organizational structures and the increased vertical integration in the different sectors of economic activity arc well-documented trends. These trends have been largely driven by technological advances, but they have also been driven by the need to bring the decision-making process closer to the customer. This was discussed in some detail by Rabeau et al. (1994).

However, one must go beyond these structural changes to look at the impact on different employee groups in certain industries. The impact of IT on employment has been an issue of great concern for policymakers. This concern was reinforced by pessimistic projections suggesting that IT will have a negative impact on overall employment. However, most empirical observations suggest that some of the negative effects of IT are offset by the redistribution of human resources and are only felt in certain types of jobs. It appears that the impacts were first felt at the secretarial and clerical levels and that middle managers and supervisors now constitute the most challenged group in many organizational contexts (Drucker 1993). Information-processing jobs, which were once the *raison d'être* of middle management, have been gradually taken over by ITs. As most members of the organization also become more "technology fluent," recent observations suggest that computer use is not merely a trend but a reality that will become even more pervasive in all organizational activities. Being computer literate is no longer a desirable quality for most workers but a necessary skill, as indicated by the major shift in the work force that has been taking place for many years (Tables 23 and 24).

This shift is taking place not only in industrialized countries but also, to a lesser extent, in the newly industrialized countries (NICs). However, the emerging trend (Tables 23 and 24) indicates that this is clearly a growing pattern

Table 23. Information work force in four developed countries.

Country	Year (%)								
	1840	1860	1880	1900	1920	1940	1960	1970	1980
United Kingdom	4.6	5.6	7.9	12.4	19.8	24.4	33.1	36.6	
United States		5.8	6.5	12.8	17.7	24.9	42.0	46.4	46.6
Australia					11.5	16.3	22.5	27.5	30.2
Germany							24.6	30.7	33.2

Source: Adapted from Kim (1994).

Table 24. Information work force in the newly industrialized countries.

Country	Year (%)		
	1960	1970	1980
Argentina	21.2	21.8	24.1
Brazil	12.0	12.2	
Mexico	10.6	16.5	20.9
Venezuela	14.1	21.3	25.6
Hong Kong	14.2	15.8	23.5
Korea	6.3	10.1	14.6
Philippines	5.8	10.5	10.8
Singapore	17.1	24.1	30.0

Source: Kim (1994).

and that NICs are gradually attaining the levels of IT adoption found in the industrialized countries. This will also have an impact on the expected skills of the ever-growing work force in the information sector.

Beyond the fundamental changes to work-force structure and employment, there are also the sociological and psychological impacts created by technology in the workplace. These can be described at four levels:

- Work-group effectiveness;
- Organizational climate;
- Job description; and
- Job satisfaction.

The impact of the introduction of IT applications on work-group effectiveness relates to changes in the attitudes of workers in terms of group work, group conflict resolution, and group respect. The second level of description captures employee perceptions of the effects of IT on organizational climate, whereas the third focuses on how IT affects the actual job being performed. Finally, the fourth considers changes in all aspects of job satisfaction, including security, challenge, and personal growth and accomplishment.

As IT increasingly penetrates all firms in all industries the basic assumption of the specialization of labour may no longer be adequate, and the capacity of the work force to change (including the ability to adapt to the pervasive sociological and psychological impacts of technology) will remain essential in helping firms compete.

Operational measures for the impacts of IT

In this section, we will propose a variety of measures to assess the impacts of IT. Let us first turn to the impact of IT on productivity. There is still considerable confusion about many aspects of productivity, including its definition (Mohanty 1992), and measuring it is therefore a difficult task.

Measurement of productivity at the macrolevel

Two groups of indicators are usually chosen at the national or sectorial level:

1.   Partial productivity index — this is the ratio of all outputs to one particular input, such as labour or capital. Although this definition makes the index easy to understand and evaluate, it also limits its use because of its restriction to only one input.

2.   Multifactor or total-factor productivity index — this takes into account all inputs, rather than only one. Even though this indicator provides a better indication of the contribution of all factors to output variation, it remains difficult to calculate.

The distinction between the two kinds of indices is very important, as shown in one recent issue of the annual report on Canadian productivity (Statistics Canada 1992). It was reported that the Canadian labour productivity index (partial indicator) steadily increased after 1975 (total increase of 25% over the 1975 level), whereas, for the same period, the multifactor productivity index remained almost the same. The following example helps one to understand how this could have happened. In a case where a firm invested substantial amounts of money in new equipment, more output per worker (or person–hour worked) could reasonably be expected. Thus, the labour index would show an increase, especially if the level of input remained the same. On the other hand, this ratio would not indicate whether the level of output had increased substantially as a function of the capital invested, but the multifactor productivity index would.

Calculation of total-factor and total-productivity indices may become quite complex, and in the last few decades a great deal of effort has been expended on developing different procedures. The measurement of productivity for manufacturing systems seems to be well documented. Some studies published in the last decade (Sumanth 1984; Sink 1985) contributed to a better understanding of the various issues involved in this difficult but essential task. Recent research seems to emphasize the measurement of productivity in white-collar industries (Drucker 1993), such as education and services. This has become important because the productivity of these industries does not compare favourably with that of manufacturing and thus contributes to lowering productivity growth at the national level (Thurow 1992).

Measurement of productivity at the microlevel

The concept of productivity has also been studied and used at a microlevel as normally understood in management science and industrial engineering, namely, at the firm or departmental level. This distinction needs to be made because, according to Carlsson (1989), most microeconomists tend to direct their analyses to the sector level rather than the firm level.

At the firm level, productivity has a very tangible meaning because it relates to what is being produced by the firm, be it services or products. Because such productivity data present a good picture of the way resources are used, they provide managers with an essential tool for planning and control. The data are also used by industrial engineers, whose task is often oriented to the optimization of operations. Comparisons of productivity levels among firms remain difficult, especially if the firms are not part of the same industry. Corrective factors must often be used to account for differences, such as in the quality of labour, wages, levels of unionization, and location. Ratios using capital as an input can also distort the comparison, depending on how expenses are accounted for.

The main difficulty continues to be a lack of comprehension of what productivity really is. Sink (1985) found that very few measures identified by managers were real productivity measures, which led him to conclude that managers still did not distinguish between productivity and certain financial ratios, such as ROI, cost per unit, and profits per sales dollar. In this context, interfirm comparison remains a rather complex task.

Measurement of productivity at the individual level

Within a narrower focus, productivity can also be studied at the group and individual levels. Although the productivity problem has been more often addressed at the national level, part of the solution lies in each individual's attitude toward his or her job, and that is of primary interest to organizational or industrial psychologists.

Many psychologically based programs, such as goal setting, work redesign, financial incentives, and work rescheduling, were found by Campbell and Campbell (1988) to be in varying degrees effective in increasing productivity. Although these authors recognized that such practices do not eliminate the need for other prescriptions at other levels, they believed that more attention should be paid to these kinds of solutions.

Table 25. Operational measures for assessing the impact of IT on productivity.

Impact of IT	Measures
Productivity at the microlevel	Multifactor or total-factor productivity index adapted to the firm (appropriate choice of inputs and outputs)
	Change in overall productivity that is due to IT applications, as perceived by the CEO
Productivity at the individual level	Ratio of appropriate outputs to hours worked by each employee
	Change in the productivity of each category of employees, as perceived by the CEO: 　　Clerical employees 　　Secretarial employees 　　Managers 　　Professionals 　　Blue-collar workers

Note: CEO, chief executive officer.

In sum, the concept of productivity is better captured at the firm and individual levels than at the national level. Table 25 shows the corresponding proposed measures.

The perceived change in productivity is less accurate than the factual measures (that is, the ratios) but is easier to compare from firm to firm. If factual measures are used, inputs and outputs are usually firm-specific and sector-specific ones (for example, outputs may be measured by the number of transactions or the number of units manufactured).

Measurement of the impact of IT on key competitive dimensions

Assessing the impact of IT applications on key competitive dimensions is a difficult but critical exercise (Kauffman and Weill 1989; Pedersen 1990). This exercise was undertaken recently in large firms by Sethi and King (1994), who provided a multidimensional index for competitive assessment of the IT applications (Table 22). Their proposed measurement is certainly valuable in the context of large companies.

In the context of SMEs, four key competitive dimensions emerge from the literature: reduced cost, increased quality, improved flexibility, and greater dependability. Each of these dimensions encompasses multiple concepts, as illustrated in Table 26.

The list of measures provided in Appendix E is as exhaustive as possible. All these measures have been tested in detailed case studies of three Canadian SMEs. Some measures can be used in any sector (for example, increased billing

Table 26. Operational measures for assessing the impact of IT on key competitive dimensions.

Impacts of IT	Measures
Cost reductions [a]	Reduction in labour costs (direct and indirect) Reduction in equipment–machinery costs Reduction in core- and support-activity costs
Quality [b]	Product Service Management Working conditions
Flexibility [c]	Processes Labour and infrastructure Design and implementation
Dependability [d]	Dependability of the equipment Delivery performance

[a] The list of potential sources of reduced costs is given in Table E1.
[b] An extensive list of quality measures is presented in Table E2.
[c] More details on the exact measures and the corresponding references are given in Table E3.
[d] Table E4 presents the detailed measures.

or order accuracy), but others are more appropriate to the services sector (for example, reduced number of complaints regarding employee courtesy) or the manufacturing sector (for example, reduced number of reworks). In SMEs, formal and accurate information, even information as basic as the number of rejects, is sometimes difficult to obtain. One may therefore wish to turn to more subjective measures, such as the relative impact of IT (from no impact to very positive impact) along these key dimensions. Lefebvre et al. (1995) offered a list of potential benefits of the adoption of IT applications, which have been extensively tested in the specific context of SMEs (see Appendix F). The list of benefits was mainly derived from the work of Miller and Roth (1988).

Measurement of the impact of IT on performance

Firm performance was defined in "Impact of IT on Performance" as comprising export performance and financial performance. The corresponding measures are shown in Table 27.

Export performance is a frequently used factual measure (Bonaccorsi 1992), whereas financial performance can be either factual or perceptual. Because CEOs of smaller independent firms are often reluctant to disclose financial data (Sapienza et al. 1988), perceptual measures for financial performance may be preferable. Such measures could include increases in sales, assets, ROI, and profits for a 3- or 5-year period relative to those of direct competitors.

Table 27. Operational measures for assessing the impact of IT on performance.

Impact of IT	Measures
Export performance	Percentage of sales realized in 　Local or regional markets 　National markets 　International markets
Financial performance	Increase during the last 3 or 5 years in 　Sales 　Assets 　ROI 　Profits

Note: ROI, return on investment.

Measurement of the impact of IT on work and employment

Table 28 captures ways to measure the level of employment and the redistribution of the work force inside firms, as well as the sociological and psychological impacts on different categories of employee.

These factors all need to be measured before and after the adoption of technology if one is to fully grasp not only the quantitative changes that have taken place in the structure and composition of the work force but also the perceptual elements that often permit one to evaluate the success of the adoption and its ultimate effects on the organizational mind set.

Table 28. Operational measures for assessing the impact of IT on work and employment.

Impacts of IT	Measures
Impact on the level of employment and the redistribution of the work force	Increase or decrease in the number of employees in the following categories (before and after the adoption of IT applications): 　Clerical employees 　Secretarial employees 　Managers 　Professionals 　Blue-collar workers
Sociological and psychological impacts on the work force	Changes in the following perceptual evaluations before and after the adoption of IT applications: [a] 　Group effectiveness 　Organizational climate 　Job description 　Job satisfaction

[a] The changes are evaluated by perceptual measures using multi-item constructs in Appendix G.

CONCLUSION

This document investigated various aspects of the adoption, diffusion, and impacts of ITs, with particular focus on the firm level. The main purpose was to identify the measures and constructs to ultimately permit the design of data-collection tools, along with identifying the main issues.

The issues presented in the document are wide ranging and cannot be covered simultaneously by a single research design. Detailed case studies would make it possible to cover a larger spectrum of issues; a survey, on the other hand, would require a focus on a smaller set of issues. Any given research strategy could combine different methodological approaches (detailed case studies, on-site interviews, and large-scale surveys, for example). In an attempt to be as exhaustive as possible, we proposed numerous measures and constructs, which can be found in the different sections of this document and its appendixes. Special attention was paid to operational measures previously tested in the context of SMEs. The formulation of these measures was grounded in the practical reality of these firms and avoided any complex or academic terminology.

Choices will therefore have to be made concerning both specific issues and measures. Data-collection tools should be adapted to the broader research objectives pursued and to the specific environments in which they are addressed. This is where the greatest challenge lies. Most researchers will agree that, beyond the identification of the pertinent dimensions to be included in a specific study, particular attention must be paid to the internal validity and reliability of these dimensions in the study environment. Furthermore, certain measures require specific research designs to realize their full potential.

Achieving fit between research objectives, research design, and specific study environments to ensure internal and external research validity and reliability will be the next step.

APPENDIX A

HARD AND SOFT TECHNOLOGIES

The following is a list (adapted from Swamidass 1994) of technologies covered by this study (a more detailed description of these technologies can be found in Appendix B):

Hard technologies
- Automated guided-vehicle systems
- Automated inspection
- Computer-aided design
- Computer-aided manufacturing, including programable automation of single or multimachine systems
- Computer-integrated manufacturing
- Computerized numerically controlled machines
- Local-area network
- Flexible manufacturing systems
- Robots

Soft technologies
- Just-in-time manufacturing
- Manufacturing cells
- Material-requirements planning
- Manufacturing-resource planning
- Statistical quality control
- Total quality management

Appendix B
Definitions of Some IT Applications

Artificial intelligence (AI)
The ability of a machine to learn from experience and perform tasks normally attributed to human intelligence, for example, problem solving, reasoning, and understanding natural language.

Automated inspection equipment for final products
Automated sensor-based equipment used for inspecting or testing final products.

Automated inspection equipment for inputs
Automated sensor-based equipment used for inspecting or testing incoming or in-process materials.

Automated storage and retrieval systems (AS–RS)
Computer-controlled equipment providing automated handling and storage of materials, parts, subassemblies, or finished products.

Automated guided-vehicle systems (AGVS)
Vehicles equipped with automated guidance devices programed to follow a path that interfaces with work stations for automated or manual loading and unloading of materials, tools, parts, or products.

Computer-aided design and engineering (CAD–CAE)
Use of computers for drawing and designing parts or products for analysis and testing of designed parts or products.

Computer-aided design for computer-aided manufacturing (CAD–CAM)
Use of CAD output for controlling machines used to manufacture the part or product.

Computer-integrated manufacturing (CIM)
Totally automated production in which all manufacturing processes are integrated and controlled by a central computer.

Computer used for control on the factory floor
Computer that may be dedicated to control on the factory floor but is capable of being reprogramed for other functions (does not include computers embedded within machines or computers used solely for data acquisition or monitoring).

Digital data representation
Use of digital representation of CAD output for controlling machines used to manufacture the part or product.

Expert system
The computerization of the knowledge of experts in narrowly defined fields, such as fault finding and designing.

Factory network
Use of local-area network (LAN) technology to exchange information between different points on the factory floor.

Flexible manufacturing cells (FMCs)
Machines with fully integrated material-handling capabilities controlled by computers or programable controllers and capable of single-path acceptance of raw material and delivery of finished product.

Flexible manufacturing systems (FMS)
Two or more machines with fully integrated material-handling capabilities controlled by computers or programable controllers and capable of single-path or multiple-path acceptance of raw material and multiple-path delivery of finished product.

Intercompany computer network (ICCN)
Use of wide-area network (WAN) technology to connect an establishment with its subcontractors, suppliers, and customers.

Manufacturing-resource planning (MRP II)
A development of MRP I for computer-based production management of machine loading and production scheduling, as well as inventory control and material handling.

Material-requirements planning (MRP I)

Computer-based production management and scheduling system to control order quantities, inventory, and finished goods.

Materials-working laser

Laser technology used for welding, sealing, treating, scribing, and marking.

Numerically controlled (NC) or computerized numerically controlled (CNC) machine

A single machine, either NC or CNC, with or without automated material-handling capabilities. NC machines are controlled by numerical command, punched on paper or plastic mylar tape, whereas CNC machines are controlled electronically through a computer residing in the machine.

Pick-and-place robot

A simple robot, with 1, 2, or 3 degrees of freedom, that transfers items from place to place by means of point-to-point moves. Little or no trajectory control is available.

Programable controller

A solid-state industrial control device that has a programable memory for storage of instructions and performs functions equivalent to those of a relay panel or wired solid-state logical control system.

Robot

A reprogramable, multifunctional manipulator designed to move materials, parts, tools, or specialized devices through variable programed motions for the performance of a variety of tasks.

Supervisory control and data acquisition (SCADA)

On-line, computer-based monitoring and control of process and plant variables of a central site.

Technical-data network

Use of local-area network (LAN) technology to exchange technical data within the design and engineering department.

APPENDIX C

MULTI-ITEM CONSTRUCTS FOR MEASURING INTERNAL FACTORS AFFECTING IT ADOPTION

The following questionnaire is designed to measure internal factors affecting the adoption of ITs.

Centralization

Which levels of management are usually responsible for making decisions of the following types?	Lower levels	Middle levels	Senior levels
1. Capital budgeting		$1-2-3-4-5-6-7$	
2. Introduction of new products		$1-2-3-4-5-6-7$	
3. Acquisition of other companies		$1-2-3-4-5-6-7$	
4. Pricing of major product lines		$1-2-3-4-5-6-7$	
5. Entry into major new markets		$1-2-3-4-5-6-7$	
6. Hiring and firing senior personnel		$1-2-3-4-5-6-7$	

Formalization

To what extent does either in the following pairs of statements reflect reality?

1. There are no written job descriptions $1-2-3-4-5-6-7$ There are complete written job descriptions for all jobs

2. Our employees make their own rules on the job $1-2-3-4-5-6-7$ Our employees must strictly abide by company rules on the job

3. We always have arguments about job overlap among our managers $1-2-3-4-5-6-7$ We never have any arguments about job overlap among our managers

Strategic orientation

To what extent do you agree or disagree with the following statements?	Strongly disagree		Strongly agree

Aggressiveness dimension

1. We sacrifice profitability to gain market share		$1-2-3-4-5-6-7$
2. We cut prices to increase market share		$1-2-3-4-5-6-7$

3. We set prices below the competition   1 — 2 — 3 — 4 — 5 — 6 — 7

4. We seek a market share position at the expense of   1 — 2 — 3 — 4 — 5 — 6 — 7
cash flow and profitability

Analysis dimension

1. We emphasize effective coordination among different   1 — 2 — 3 — 4 — 5 — 6 — 7
functional areas

2. We believe that information systems provide support   1 — 2 — 3 — 4 — 5 — 6 — 7
for decision-making

3. When confronted with a major decision, we usually try   1 — 2 — 3 — 4 — 5 — 6 — 7
to develop a thorough analysis

4. We encourage the use of planning techniques   1 — 2 — 3 — 4 — 5 — 6 — 7

5. We encourage the use of the output of management   1 — 2 — 3 — 4 — 5 — 6 — 7
information and control systems

6. We encourage human-resource power planning and   1 — 2 — 3 — 4 — 5 — 6 — 7
performance appraisal of senior managers

Defensiveness dimension

1. We have made significant modifications to manu-   1 — 2 — 3 — 4 — 5 — 6 — 7
facturing technology

2. We encourage the use of cost-control systems for   1 — 2 — 3 — 4 — 5 — 6 — 7
monitoring performance

3. We encourage the use of production-management   1 — 2 — 3 — 4 — 5 — 6 — 7
techniques

4. We emphasize product quality (for example, through   1 — 2 — 3 — 4 — 5 — 6 — 7
the use of quality circles)

Futurity dimension

1. Our criteria for resource allocation generally reflect   1 — 2 — 3 — 4 — 5 — 6 — 7
short-term considerations (reversed scale)

2. We emphasize basic research to provide us with a   1 — 2 — 3 — 4 — 5 — 6 — 7
future competitive edge

3. We forecast key indicators of operations   1 — 2 — 3 — 4 — 5 — 6 — 7

4. We try to obtain a formal tracking of general trends   1 — 2 — 3 — 4 — 5 — 6 — 7

5. We analyze critical issues with "what if"   1 — 2 — 3 — 4 — 5 — 6 — 7

Proactiveness dimension

1. We are constantly seeking new opportunities related to   1 — 2 — 3 — 4 — 5 — 6 — 7
present operations

2. We are usually the first ones to introduce new brands   1 — 2 — 3 — 4 — 5 — 6 — 7
or products on the market

3. We are constantly on the lookout for business that can   1 — 2 — 3 — 4 — 5 — 6 — 7
be acquired

4. Competitors generally preempt us by expanding   1 — 2 — 3 — 4 — 5 — 6 — 7
capacity (reversed scale)

5. Our operations in later stages of life cycle are   1 — 2 — 3 — 4 — 5 — 6 — 7
strategically diminished

Riskiness dimension

1. Our activities can be generally characterized as high   1 — 2 — 3 — 4 — 5 — 6 — 7
    risk

2. We adopt a rather conservative view when making   1 — 2 — 3 — 4 — 5 — 6 — 7
    major decisions (reversed scale)

3. Our new projects are approved on a stage-by-stage   1 — 2 — 3 — 4 — 5 — 6 — 7
    basis, rather than with blanket approval (reversed
    scale)

4. We tend to support projects where the expected   1 — 2 — 3 — 4 — 5 — 6 — 7
    returns are certain (reversed scale)

5. Our operations have generally followed the tried-and-   1 — 2 — 3 — 4 — 5 — 6 — 7
    true paths (reversed scale)

Technology policy

To what extent do you agree or disagree with the   Strongly   Strongly
following statements?   disagree   agree

1. The policy of this firm has always been to explore the   1 — 2 — 3 — 4 — 5 — 6 — 7
    most up-to-date production (operations) technology

2. We are going ahead with plans to evaluate new   1 — 2 — 3 — 4 — 5 — 6 — 7
    processing equipment

3. We have a long tradition and reputation in our industry   1 — 2 — 3 — 4 — 5 — 6 — 7
    for attempting to be first to try out new methods and
    equipment

4. We plan to increase our spending on research and   1 — 2 — 3 — 4 — 5 — 6 — 7
    development over the next 5 years

5. We spend more than most firms in our industry on new-   1 — 2 — 3 — 4 — 5 — 6 — 7
    product development

6. We are actively engaged in a campaign to recruit the   1 — 2 — 3 — 4 — 5 — 6 — 7
    best qualified technical personnel (engineering and
    production)

7. We are actively engaged in a campaign to recruit the   1 — 2 — 3 — 4 — 5 — 6 — 7
    best qualified marketing personnel

8. We are one of the few firms in our industry that does   1 — 2 — 3 — 4 — 5 — 6 — 7
    technological forecasting for products

9. We are one of the few firms in our industry that does   1 — 2 — 3 — 4 — 5 — 6 — 7
    technological forecasting for production processes

Technological awareness

Gauge your response to each of the following   Minimally   Somewhat   Very
questions

1. Are you aware of the most recent technological   1 — 2 — 3 — 4 — 5 — 6 — 7
    developments?

2. Are you up to date on the availability of the most   1 — 2 — 3 — 4 — 5 — 6 — 7
    recent technological developments on the market?

3. Are you aware of the comparative advantages that   1 — 2 — 3 — 4 — 5 — 6 — 7
    you can get from these most recent developments?

APPENDIX D

DETAILED MEASUREMENT OF SOME ELEMENTS OF THE DECISION-MAKING PROCESS FOR IT ADOPTION

The following questionnaire is designed to measure attitudes toward risk as an element of the decision process in the adoption of ITs.

Attitudes toward risk

Which do you favour?

1. Low-risk projects (with a guaranteed but moderate return on investment)	1 — 2 — 3 — 4 — 5 — 6 — 7	High-risk projects (with the chance of a high return on investment)
2. Gradual and moderate reactions to outside changes	1 — 2 — 3 — 4 — 5 — 6 — 7	Aggressive and far-reaching reactions to outside changes

Which do you prefer?

1. To introduce changes after competitors	1 — 2 — 3 — 4 — 5 — 6 — 7	To introduce changes before competitors
2. Time-tested methods	1 — 2 — 3 — 4 — 5 — 6 — 7	Innovation

Top-down–bottom-up decision-making process

A firm can be said to have a top-down decision-making process if the relative importance of the chief executive officer's (CEO's) influence (on both initiating and adopting IT applications) is much stronger than the relative importance of functional groups' influence (research and development, marketing, production, accounting–finance, and IT groups). Depending on the group of firms, the CEO's influence should be greater than the average sum of all influences from functional groups plus 1, 2, or 3 standard deviations.

A firm has a bottom-up decision-making process if the CEO's influence is not greater than that of the functional group.

APPENDIX E

MEASURES FOR ASSESSING THE IMPACT OF ITS ON KEY COMPETITIVE DIMENSIONS

Several measures can be used to measure the impact of ITs on key competitive dimensions.

Cost reductions can occur in stock, planning, follow-up, etc. and can be assessed in monetary terms (Table E1).

Table E1. Cost reductions.

Variable	Measure	References
Stock	Monetary	Primrose (1990) Swann and O'Keefe (1990) Troxler (1990)
Planning and follow-up	Monetary	Meredith (1988) Noori (1990)
Handling	Monetary	Meredith (1988) Noori (1990) Troxler (1990) Pasewack (1991)
Equipment — maintenance	Monetary	Meredith (1988) Troxler (1990)
Equipment — repair	Monetary	Troxler (1990)
Direct labour	Monetary	Troxler (1990)
Indirect labour	Monetary	Noble (1990)
Implementation	Monetary	Noble (1990) Noori (1990) Troxler (1990)
Waste	Monetary	Meredith (1988) Noble (1990) Noori (1990) Primrose (1990)

(continued)

Table E1 continued.

Variable	Measure	References
Rework	Monetary	Meredith (1988) Noble (1990) Noori (1990) Primrose (1990)
Tool	Monetary	Noori (1990) Troxler (1990)
Inspection	Monetary	Meredith (1988) Troxler (1990)
Raw materials (waste)	Monetary	Meredith (1988) Noble (1990) Swann and O'Keefe (1990)
Down time	Monetary	Primrose (1990) Swann and O'Keefe (1990)
Transportation	Monetary	Primrose (1990)
Space savings	Monetary	Meredith (1988) Noble (1990) Noori (1990) Primrose (1990) Swann and O'Keefe (1990)
Sales or customers lost	Monetary	Swann and O'Keefe (1990)
New customer requests	Monetary	Clemmer (1990)
Order	Monetary	Noori (1990) Troxler (1990)
Warranty	Monetary	Primrose (1990) Pasewack (1991)
After-sale service cost	Monetary	Primrose (1990) Pasewack (1991)
Design	Monetary	Noble (1990) Noori (1990) Troxler (1990)
Product modification	Monetary	Meredith (1988) Noori (1990) Troxler (1990)
Financing receivables	Monetary	Meredith (1988) Noori (1990) Primrose (1990) Troxler (1990)
Training and recruiting	Monetary	Meredith (1988) Noori (1990) Primrose (1990) Troxler (1990)

(continued)

Table E1 concluded.

Variable	Measure	References
Information input	Monetary	Swann and O'Keefe (1990)
Processing errors	Monetary	Primrose and Leonard (1986a) Primrose (1990)
Training time	Monetary	Swann and O'Keefe (1990)
Number of working days lost (work accidents)	Monetary	Swann and O'Keefe (1990)

Quality is multidimensional concept. IT applications can improve the quality of the product, customer service, management and functional activities, and working conditions (Table E2).

Table E2. Improved quality.

Variable	Measure	References
Product		
Performance	Speed of processing	Juran and Mirano (1980) Quarante (1984) Mizuno (1988)
	Level of product precision	Juran and Mirano (1980) Quarante (1984) Mizuno (1988)
	Level of consumption of resources (energy) for the customer	Quarante (1984) Mizuno (1988) Hall et al. (1990)
	Simplicity of use	Mizuno (1988)
	Number of options in comparison with competing products	Quarante (1984)
Aesthetics	Shapes and proportions of the product	Quarante (1984)
	Look of the product	Quarante (1984)
	Product dimensions	Quarante (1984)
Conformance	Defect rate	Hall et al. (1990)
	Number of interventions (guaranteed)	Garvin (1988) Bartezzaghi and Turco (1989)

(continued)

Table E2 continued.

Variable	Measure	References
Conformance (continued)	First-inspection pass rate	Hall et al. (1990)
	Last-inspection pass rate	Hall et al. (1990)
	Number of interventions (not guaranteed)	Bartezzaghi and Turco (1989)
Durability and maintainability	Mean period between maintenance	Juran and Grynia (1980)
	Mean time to repair	Juran and Grynia (1980)
	Mean time to preventive maintenance	Hall et al. (1990)
	Product life expectancy	Quarante (1984) Juran (1988)
	Mean time for breakdowns	Juran and Grynia (1980)
	Percentage of available time (mean time between failures divided by the mean time between failures plus mean time for breakdowns)	Juran and Grynia (1980) Hall et al. (1990) Bartezzaghi et al. (1992)
	Ratio of maintenance hours to hours of use	Juran and Grynia (1980)
Customer service		
Technical assistance	Mean time between the call and the intervention	Juran and Grynia (1980) Bartezzaghi and Turco (1989)
	Number of service calls per 100 guaranteed units	Juran and Grynia (1980)
	Mean time between service calls	Juran and Grynia (1980)
Serviceability	Percentage of repairs correctly done on the first service call	Hall et al. (1990)
	Number of repairs not done on the first attempt	Garvin (1988) Juran (1988)
	Total number of customer complaints	Juran and Grynia (1980)
	Number of complaints by customers regarding the employees' courtesy	Juran (1988)
	Mean delay time to perform a repair	Juran and Grynia (1980) Garvin (1988) Bartezzaghi and Turco (1989)

(continued)

Table E2 continued.

Variable	Measure	References
Serviceability (continued)	Time to resolve customer complaints	Clemmer (1990)
Management and functional activities		
Administration	Inventory turnover	Hall et al. (1990)
	Percentage of documents containing errors	Juran (1988)
	Billing accuracy (error rate)	Hall et al. (1990)
Marketing and sales	Order accuracy (error rate)	Hall et al. (1990)
	Average customer lead time	Hall et al. (1990)
	Number of canceled customer orders	Juran (1988)
	Number of complaints related to product design	Harrington (1991)
	Customer turnover	
	Mean time between customer order and delivery	Bartezzaghi and Turco (1989)
	Mean delivery delay (between end of production and shipping)	Bartezzaghi and Turco (1989)
Purchasing	Number of delays for a part shortage	Hall et al. (1990)
	Percentage of stock shortage	Juran (1988)
	Longest supplier lead time	Hall et al. (1990)
	Cost related to acquisition of poor-quality product	Juran (1988)
	Percentage of repeated orders related to poor-quality product received the first time	Juran (1988)
	Percentage of defective parts returned to the supplier	Hall et al. (1990)
	Rate of on-time arrival	Hall et al. (1990)
Production	Time to build	Hall et al. (1990)
	Unplanned equipment down time	Hall et al. (1990)
	Time devoted to improving the operation's quality	Asher (1988)

(continued)

Table E2 concluded.

Variable	Measure	References
Production (continued)	Average time between customer request and production start-up	Bartezzaghi and Turco (1989)
	Number of meetings on quality improvement	Asher (1988)
Human resources	Time spent on training and recycling	Asher (1988)
	Average time required to fill vacant jobs	
Engineering	Design cost	Bartezzaghi et al. (1992) Hall et al. (1990)
	Number of parts	Hall et al. (1990)
	Drawing accuracy (error rate)	Hall et al. (1990)
	Time to make changes on plans	Hall et al. (1990)
	Number of engineering changes per drawing	Harrington (1991)
	Time spent on new-product and new-process development	Asher (1988)
Working conditions	Rate of absenteeism	Swann and O'Keefe (1990)
	Rate of employee turnover	Swann and O'Keefe (1990)
	Number of grievances	Swann and O'Keefe (1990)

Flexibility can be measured with respect to processes, labour and infrastructure, and design and implementation (Table E3).

Table E3. Flexibility.

Variable	Measure	References
Processes		
Machine	Number of different operations	Carter (1986) Brill and Mandelbaum (1987)
	Time required to switch from one operation to another	Browne et al. (1984) Carter (1986)
Routing	Robustness of the system when breakdowns occurs	Browne et al. (1984)
	Percentage of loss of productivity that is due to change in product mix	Carter (1986)
Production	Size of the universe of parts the system is capable of producing	Bartezzaghi et al. (1992)
Operation	Number of different processing plans for fabrication	Sethi and Sethi (1990)
Processing	Extent to which the product mix can be changed while maintaining efficient production	Sethi and Sethi (1990)
Expansion	Investment cost required to double the production capacity (ratio of cost of equipment needed to the current equipment cost)	Carter (1986) Sethi and Sethi (1990)
Volume	Ability to operate profitably at different overall output levels	Browne et al. (1984) Bartezzaghi and Turco (1989)
Product	Time required to switch from one part mix to another	Buzacott (1982) Browne et al. (1984)
Labour and infrastructure		
Numerical	Capacity to adjust work force to sales fluctuation	Atkinson (1985)
Functional	Capacity to dispatch work force as function of production need	Atkinson (1985)
Training	Average number of abilities or tasks mastered by production employees	Bartezzaghi and Turco (1989) Hall et al. (1990)
	Work force mobility	Bartezzaghi et al. (1992)
Finance	Capacity to secure the work force by offering good working conditions	Atkinson (1985)

(continued)

Table E3 concluded.

Variable	Measure	References
Design and implementation		
Modification	Required time between product change and introduction of the production line	Noori (1990)
Innovation	Average number of new products introduced per year	Gerwin (1987) Maskell (1991)
	Mean period between two consecutive innovations	Azzone et al. (1991)
	Percentage of products and models new within 2 years	Hall et al. (1990)
	Rate of new-product introduction	Stalk and Hout (1990)
New-product design	Time required for the design of a new product	Azzone et al. (1991)
	Degree of customization of products	Bartezzaghi and Turco (1989)
	Development time of a new model (from concept meeting to first unit produced)	Hall et al. (1990) Bartezzaghi et al. (1992)
	Percentage of standard parts included in the design of a new product	Bartezzaghi et al. (1992)
Market	Time from customer recognition of need to delivery	Stalk and Hout (1990)
	Time from idea to market	Stalk and Hout (1990) Maskell (1991)
	Time between the decision to introduce a new product or variation and production start-up	Bartezzaghi and Turco (1989)

Improved dependability can be expressed in terms of equipment and delivery performance (Table E4).

Table E4. Dependability.

Variable	Measure	References
Equipment	Mean time between failures	Garvin (1988) Juran (1988) Bartezzaghi and Turco (1989)
	Mean time before first mechanical breakdown	Juran and Grynia (1980) Garvin (1988)
	Frequency of failures	Juran (1988)
	Percentage of new units returned as defective	Hall et al. (1990)
	Ratio of repairs to 100 running hours	Juran and Grynia (1980)
Delivery performance	Percentage of requests answered on time	Harrington (1991) Oakland and Wynne (1991)
	Percentage of on-time deliveries	Hall et al. (1990) Stalk and Hout (1990) Azzone et al. (1991) Oakland and Wynne (1991)
	Required time to correct a problem	Harrington (1991)
	Time to respond to a customer request	Bartezzaghi and Turco (1989)

APPENDIX F
ALTERNATIVE MEASURES FOR ASSESSING THE IMPACT OF ITS

The relative impact of ITs can be assessed using the following 24 potential benefits, mainly derived from the work of Miller and Roth (1988). The benefits can be measured using seven-point Likert scales. These measures have been extensively tested in the context of small and medium-sized enterprises (Lefebvre et al. 1995).

- Space reduction;
- Reduction in inventory levels;
- Increased use of machinery and equipment;
- Reduction in capital investment (for example, equipment, machinery);
- Increase in productivity of production employees;
- Increase in productivity of nonproduction employees;
- Decrease in set-up time;
- Reduction in rate of rejected items;
- Decrease in rate of production of defective items;
- Increase in flexibility of manufacturing process;
- Reduction in lead time;
- Reduction of managerial controls;
- Improvement of working conditions;
- Improvement of the firm's image in the market;
- Increase in number of customized products offered;
- Increase in variety of products offered;
- Increase in number of new products offered;
- Increase in the durability of products offered;
- Increase in the reliability of products offered;
- Decrease in the number of complaints by clients;
- Decrease in the number of repairs on products sold;

- Decrease in production costs (manufacturing);
- Decrease in cost of products; and
- Ability to meet deadlines.

APPENDIX G

MEASURES FOR ASSESSING THE SOCIOLOGICAL AND PSYCHOLOGICAL IMPACTS OF ITS ON THE WORK FORCE

The sociological and psychological impacts of ITs on the work force can be assessed using the multi-item constructs developed by Groebner and Merz (1994) (Table G1).

Table G1. Sociological and psychological impacts.

Impact	Measure
Group effectiveness	Goal clarity
	Individual participation
	Acceptance of feelings
	Problem diagnosis
	Leadership
	Decision-making
	Mutual trust
	Creativity and growth
Organizational climate	Conformity required
	Personal responsibility
	Standards
	Rewards given
	Division clarity
	Warmth and support
	Leadership accepted
Job description	Required to work closely with others
	Autonomy in your job

(continued)

Table G1 concluded.

Impact	Measure
Job description (continued)	Doing a whole piece of work
	Variety in your job
	Feedback from job itself
Job satisfaction	Job security
	Pay and fringe benefits
	Personal growth
	Co-workers
	Respect from boss
	Accomplishment
	Meeting others
	Support from boss
	Fair pay
	Independence
	Helping others
	Challenge
	Quality of supervision

BIBLIOGRAPHY

These references are classified according to the chapters of this document and also correspond more or less to the sections of Chapter 4.

Chapter 1. Typologies and Rates of Diffusion

ABS (Australian Bureau of Statistics). 1989. Manufacturing technology statistics, Australia, 30 June 1988, summary. ABS, Canberra, ACT, Australia. Catalogue No. 8123.0.

——— 1990. Survey of manufacturing technology, 1988 — international comparison: survey of manufacturing technology, Australia 1988 and survey of manufacturing technology, Canada, 1989. ABS, Canberra, ACT, Australia. Working Paper.

——— 1990. Survey of manufacturing technology: 1988, main tables and employment tables, Australia, 1990. ABS, Canberra, ACT, Australia. Working Paper.

Allen, T.J.; Scott Morton, M.S. 1994. Information technology and the corporation of 1990s. Oxford University Press, New York, NY, USA.

Barrera, M. 1986. Diffusin de la tecnologia computacional en una economa abierta: el caso de Chile. Centro de Estudios Sociales, Santiago., Chile.

Burch, J. 1989. EDI: the demise of paper. Information Executive, p. 52.

Business Week. 1994. Nov, p. 82.

——— 1995. Annual report on IT. Jun, pp. 45–67.

Dedrick, J.L.; Goodman, S.E.; Kraemer, L.L. 1995. Little engines that could: computing in small energetic countries. Communications of the ACM, 38(5), 21–26.

Ducharme, l.M.; Gault, F. 1992. Surveys of advanced manufacturing technology. Science and Public Policy, Dec, pp. 393–399.

Dunne, T.H. 1991. Technology usage in US manufacturing industries: new evidence from the survey of manufacturing technology. Center for Economic Studies, US Bureau of the Census. Discussion Paper 91–7.

Economist, The. 1994. Between two worlds: a survey of manufacturing technology. 5 Mar, pp. 1–18.

Edquist, C.; Jacobson, S. 1988. Flexible automation: the global diffusion of new technology in the engineering industry. Basil Blackwell, Oxford, U.K.

Florida, R. 1991. The new industrial revolution. Futures, 23(6), 559–576.

Ford, F.N.; Ledbetter, W.N.; Gaber, B.S. 1985. The evolving factory of the future: integrating manufacturing and information systems. Information and Management, 8, 75–80.

Fortier, Y.; Ducharme, I.M.; Gault, F. 1993. A comparison of the use of advanced manufacturing technologies in Canada and the United States. STI Review, No. 12, 81–100.

Gerwin, D.; Kolodny, H. 1992. Management of advanced manufacturing technology — strategy, organization and innovation. John Wiley & Sons Inc., New York, NY.

Gray, A.E.; Seidmann, A.; Stecke, K.E. 1993. A synthesis of decision models for tool management in automated manufacturing. Management Science, 39(5), 549–567.

Hsu, C.; Skevington, C. 1987. Integration of data and knowledge in manufacturing enterprises: a conceptual framework. Manufacturing Systems, 6, 227–285.

ISTC (Industry, Science and Technology Canada). 1990. Technologies in services, report for Communications Canada and Industry, Science and Technology Canada. ISTC, Ottawa, ON, Canada. Catalogue No. C2–133/1990.

Jaakkola, H.; Tenhunen, H. 1993. The impact of information technology on Finnish industry: a review of two surveys. STI Review, No. 12, 53–80.

Jang, S.L.; Norsworthy, J.R. 1992. Technology adoption in US manufacturing: a comparison of foreign associated and domestic plants. Center for Science and Technology Policy, Rensselaer Polytechnic Institution, Troy, NY, USA. Working Paper 92–1.

Julien, P.A.; Carrière, J.B.; Hébert, L. 1988. La Diffusion des nouvelles technologies dans trois secteurs industriels. Conseil de la Science et de la Technologie, Government of Quebec, Québec, QC, Canada.

Kelley, M.R.; Brooks, H. 1988. The state of computerized automation in us manufacturing. Center for Business and Government, Harvard University. Report.

Lefebvre, É.; Lefebvre, L.A. 1992. CEO characteristics and technology adoption in smaller manufacturing firms. Journal of Engineering and Technology Management, 9, 243–277.

———— 1993. Domaines privilégiés d'application de la micro-informatique : la bureautique, la télématique et la robotique. Micro-informatique, entreprises et société, Télé Université, pp. 34–83.

Mansfield, E. 1993. The diffusion of flexibility manufacturing systems in Japan, Europe and the United States. Management Science, 39(2).

McFetridge, D.G. 1992. Analysis of recent evidence on the use of advanced technologies in Canada. Carleton University, Ottawa, ON, Canada.

Moad, J. 1989. Navigating cross-functional IS waters. Datamation, Mar.

Mori, S. 1993. Diffusion of advanced manufacturing systems in Japan. STI Review, No. 12, 101–124.

Nijhowne, S. et al. 1991. Concordance between the standard industrial classifications of Canada and the United States, 1980 Canadian SIC — 1987 United States SIC. Statistics Canada, Ottawa, ON, Canada. Catalogue No. 12–574.

Northcott, J.; Vickery, G. 1993. Surveys of the diffusion of microelectronics and advanced manufacturing technology. STI Review, No. 12, 7–35.

O'Brien, J.A. 1990. Management information systems: a managerial end user perspective. Irwin, Boston, MA, USA.

O'Brien, J.A. 1993. Management information systems: a managerial end user perspective (2nd ed.). Irwin, Boston, MA, USA.

OECD (Organization for Economic Co-operation and Development). 1990. Description of innovation surveys and surveys of technology use carried out in OECD member countries. OECD, Paris, France.

——— 1992. OECD proposed guidelines for collecting and interpreting technological innovation data: Oslo manual. OECD, Paris, France.

Pennings, J.M. 1987. On the nature of new technology as organizational innovation. In Pennings, J.M.; Buitendam, A., ed., New technology as organizational innovation. Ballinger Publishing Company, Cambridge, MA, USA.

Schultz-Wild, R. 1991. CIM and future factory structures in Germany. Futures, 23(10), 1032–1046.

Statistics Canada. 1988. Survey of manufacturing technology 1987. Statistics Canada, Ottawa, ON, Canada. Final Report.

——— 1989. Survey of manufacturing technology 1989: statistical tables. Services, Science and Technology Division, Statistics Canada, Ottawa, ON, Canada. Working Paper ST–89–10.

——— 1990. The Statistics Canada business register. Systems Development Division, Statistics Canada, Ottawa, On, Canada. Aug.

——— 1991. Survey of manufacturing technology 1989, indicators of science and technology 1989. Statistics Canada, Ottawa, ON, Canada. Catalogue No. 88–002, Vol. 1, No. 4.

——— 1991. Survey of manufacturing technology 1989: analysis at the 3–digit SIC level, indicators of science and technology 1990. Statistics Canada, Ottawa, ON, Canada. Catalogue No. 88–002, Vol. 2, No. 3.

Statistics Canada; Baldwin, J.; Sabourin, D. 1995. Technology adoption in Canadian manufacturing. Statistics Canada, Ottawa, ON, Canada. Catalogue No. 88–512.

Swamidass, P.M. 1994. Technology on the factory floor: benchmarking manufacturing technology use in the United States. The Manufacturing Institute. Dec.

Thacker, R.M. 1989. A new CIM model: a blueprint for the computer-integrated manufacturing enterprise. SME, Dearborn, MI, USA.

Thurow, L.C. 1987. A weakness in process technology. Science, 238, 1659–1663.

US Bureau of the Census. 1989. Manufacturing technology 1988. US Bureau of the Census, Department of Commerce, Washington, DC, USA. Current Industrial Reports, SMT (88)–1.

Wilder, C. 1989. Knocking down the organizational walls. Computerworld, Apr.

Yazici, H.; Benjamin, C.; McGlauglin, J. 1994. AI–based generation of production engineering labor standards. IEEE Transactions on Engineering Management, 41(3), 302.

Chapter 2. Factors Affecting Adoption

Abrahamson, E. 1991. Managerial fads and fashions: the diffusion and rejection of innovations. Academy of Management Review, 16(3), 586–612.

Adler, P.S. 1989. CAD/CAM: managerial challenges and research issues. IEEE Transactions on Engineering Management, 36, 202–215.

———— 1989. Technology strategy: a guide to the literature. In Rosenbloom, R.S.; Burgelman, R.A., ed., Research on technological innovation, management and policy. Vol. 4. JAI Press Inc., Greenwich, CT, USA. pp. 25–151.

Adler, P.S.; Shenhar, A. 1990. Adapting your technological base: the organizational challenge. Sloan Management Review, Fall, 25–37.

Alexander, M.B.; Walasa, C. 1992. Implementing emerging information technologies: the example of imaging information systems (abstract). Proceedings, 13th International Conference on Information Systems, Dec, Dallas, TX, USA.

Alpar, P.; Ein–Dor, O. 1991. Major IS concerns of entrepreneurial organizations. Information and Management, 20, 1–11.

Amoako, K.; Gyampah; Maffei, M.J. 1989. The adoption of flexible manufacturing systems: strategic considerations. Technovation, 9, 479–491.

Angell, I.O.; Smithson, S. 1991. Information systems management: opportunities and risks. Macmillan, Basingstoke, UK.

Arnold, E. 1987. Some lessons from government information technology policies. Technovation, 5(4), 247–268.

Baldwin, J.; Chandler, W.; Le, C.; Papailiadis, T. 1994. Strategies for success. Statistics Canada, Ottawa, ON, Canada. Catalogue No. 61–523E.

Ball, L.D.; Dambolena, I.G.; Hennessey, H.D. 1987. Identifying early adopters of large software systems. Data Base, pp. 21–27.

Bikson, T.K.; Gutek, R.A.; Mankin, D.A. 1987. Implementing computerized procedures in office settings. The Rand Corporation, Santa Monica, CA, USA. Report No. Rand/R–3077–NSF/IRIS.

Boddy, D.; Buchanan, D.A. 1986. Managing new technology. Basil Blackwell, Oxford, UK.

Boone, M.E. 1990. Leadership and the computer. Prima Publishing, Rocklin, CA, USA.

Boynton, A.C.; Zmud, R.W.; Jacobs, G.C. 1994. The Influence of IT management practice on it use in large organizations. MIS Quarterly, 18(3), 299–318.

Brancheau, J.C.; Wetherbe, J.C. 1990. The adoption of spreadsheet software: testing innovation diffusion theory in the context of end–user computing. Information Systems Research, 1, 115–143.

Brown, A.D.; Roberts, H. 1992. Implementing information systems: some practical advice and a richer model. Creativity and Innovation Management, 1(3), 121–126.

Brown, C.V.; Magill, S.L. 1994. Alignment of the IS functions with the enterprise: toward a model of antecedents. MIS Quarterly, 18(4), 371–403.

Burgelman, R.A.; Rosenbloom, R.S. 1989. Technology strategy: an evolutionary process perspective. In Rosenbloom, R.S.; Burgelman, R.A., ed., Research on technological innovation, management and policy. Vol. 4. JAI Press Inc., Greenwich, CT, USA.

Burrows, P. 1994. Giant killers on the loose. Business Week, No. 3372, 18 May, pp. 108–110.

Carnoy, M. 1994. The new global economy, information technology and restructuring education. International Journal of Technology Management, 9(3/4), 270–286.

Clark, A. 1987. Small business computer systems. Hodder and Stoughton, London, UK.

Clark, K.B.; Hayes, R.H. 1987. Exploring factors affecting innovation and productivity growth within the business unit. In Clark, K.B.; Hayes, R.H.; Lorenz, C., ed., The uneasy alliance. Harvard Business School Press, Boston, MA, USA.

Collins, P.D.; Hage, J.; Hull, F.M. 1988. Organizational and technological predictors of change in automaticity. Academy of Management Journal, 31, 512–543.

Cooley, P.L.; Walz, D.T.; Walz, D.B. 1987. A research agenda for computers and small business. American Journal of Small Business, 11(3), 31–42.

Cooper, R.B.; Zmud, R.W. 1990. Information technology implementation research: a technological diffusion approach. Management Science, 36, 123–139.

Cragg, P.B.; King, M. 1993. Small-firm computing: motivators and inhibitors. MIS Quarterly, 17(1), 47–60.

Daly, A.D.M.; Hitchens, W.N.; Wagner, K. 1985. Productivity, machinery and skills in a sample of British and German manufacturing plants — results from a pilot study. National Institute Economic Review, Feb, pp. 48–61.

Davenport, T.H. 1994. Saving IT's soul: human-centered information management. Harvard Business Review, Mar/Apr, 119–131.

Davis, F. 1989. Perceived usefulness, perceived ease of use, and user acceptance of information technology. MIS Quarterly, 13(3), 319–340.

Davis, F.; Bagozzi, R.; Warshaw, R. 1989. User acceptance of computer technology: a comparison of two theoretical models. Management Science, 35, 982–1003.

Delone, W.H. 1988. Determinants of success for computer usage in small business. MIS Quarterly, 12(1), 51–61.

Doukidis, G.I.; Smithson, S.; Lybereas, T. 1993. Approaches to computerization in small businesses in Greece. Proceedings, 13th International Conference on Information Systems, Dec, Dallas, TX, USA.

Driggers, L. 1993. Finding a system for a small agency. American Agent and Broker, 65(6), 40–44.

Earl, M.J. 1993. Experience in strategic information systems planning. MIS Quarterly, 17(1), 1–24.

Elango, B.; Meinhart, W.A. 1994. Selecting a flexible manufacturing system: a strategic approach. Long Range Planning, 27(3), 118–126.

Ettlie, J.E.; Bridges, W.P. 1987. Technology policy and innovation in organizations in new technology as innovation. In Pennings, J.M.; Buitendam, A., ed., New technology as organizational innovation. Ballinger Publishing Company, Cambridge, MA, USA.

Farhoomand, F.; Heryeyk, G.P. 1985. The feasibility of computers in small business environments. American Journal of Small Business, 9(4), 15–22.

Fitchman, R.G. 1992. Information technology diffusion: a review of empirical research. Proceedings, 13th International Conference on Information Systems, Dec, Dallas, TX, USA.

Fletcher, K.; Wheeler, C.; Wright, J. 1994. Strategic implementation of database marketing: problems and pitfalls. Long Range Planning, 27(Feb), 133–143.

Gable, G.G. 1991. Consultant engagement for computer systems selection: a pro–active client role in small businesses. Information and Management, 20, 83–93.

Gable, G.G.; Raman, K.S. 1992. Government initiatives for IT adoption in small businesses. International Information Systems, 1(1), 68–93.

Garvin, D.A. 1993. Building a learning organization. Harvard Business Review, Jul/Aug, 78–91.

Gatignon, H.; Robertson, T.S. 1989. Technology diffusion: an empirical test of competitive effects. Journal of Marketing, 53, 35–49.

Geipel, G.L. 1991. The failure and future of information technology policies in Eastern Europe. Technology in Society, 13(1/2), 207–228.

Geisler, E. 1992. Managing information technologies in small business: some practical lessons and guidelines. Journal of General Management, 18(1), 74–81.

Gerwin, D.; Kolodny, H. 1992. Management of advanced manufacturing technology — strategy, organization and innovation. John Wiley & Sons Inc., New York, NY, USA.

Gillies, J. 1994. EDI — who pays? CMA Magazine, Dec/Jan, p. 5.

Gurbaxani, V. 1990. Diffusion in computing networks. Communications of the ACM, 33(12), 65–75.

Gurbaxani, V.; Mendelson, H. 1990. An integrative model of information systems spending growth. Information Systems Research, 1, 23–46.

Heikkila, J.; Saarinen, T.; Saaksjarvi, M. 1991. Success of software packages in small business: an exploratory study. European Journal of Information Systems, 1(3), 159–169.

Hottenstein, M.P.; Dean, J.W., Jr. 1992. Managing risk in advanced manufacturing technology. California Management Review, 34, 122–126.

Howard, C.; Vitalari, N. 1992. An empirical examination of the determinants of electronic communications technology use and its impact on supervisor–subordinate interactions. Proceedings, 13th International Conference on Information Systems, Dec, Dallas, TX, USA.

Howell, J.M.; Higgins, C.A. 1990. Champions and technological innovation. Administrative Science Quarterly, 35, 317–341.

Huff, S.L.; Munro, M.C. 1989. Managing micro proliferation. Journal of Information Systems Management, 72–75.

Jackson, W.M.; Palvia, P. 1991. The state of computers and MIS in small business and some predictive measures of computer use. Journal of Computer Information Systems, 31(Spring).

Kimberly, J.R. 1987. Organizational and contextual influences on the diffusion of technological innovation. In Pennings, J.M.; Buitendam, A., ed., New technology as organizational innovation. Ballinger Publishing Company, Cambridge, MA, USA.

Kirby, D.A.; Turner, J.S. 1993. IT and the small retail business. International Journal of Retail and Distribution Management, 21(7), 20–27.

Kwon, T.H. 1990. A diffusion of innovation approach to MIS diffusion: conceptualization, methodology, and management strategy. Proceedings, 11th International Conference on Information Systems, Dec, Copenhagen, Denmark. pp. 139–146.

Lees, J.D.; Lees, A.D. 1987. Realities of small business information systems implementation. Journal of Systems Management, Jan, 6–13.

Lefebvre, É.; Lefebvre, L.A. 1992. CEO characteristics and technology adoption in smaller manufacturing firms. Journal of Engineering and Technology Management, 9, 243–277.

Lefebvre, É.; Lefebvre, L.A.; Harvey, J. 1993. Competing internationally through multiple innovative efforts. R&D Management, 23(3), 227–237.

Lefebvre, É.; Lefebvre, L.A.; Roy, M.J. 1995. Technological penetration and organizational learning: the cumulative effects. Technovation, 15(8), 511–522.

Lefebvre, L.A.; Harvey, J.; Lefebvre, É. 1991. Technological experience and the technology adoption decision in small manufacturing firms. R&D Management, 21, 241–249.

Lefebvre, L.A.; Lefebvre, É.; Ducharme, J. 1989. Introduction and use of computers in small business: a study of the perceptions and expectations of managers. Canada Department of Communications, Ottawa, ON, Canada. MCC–CWARC–DLR–85/6–009. 105 pp.

Lefebvre, L.A.; Lefebvre, É.; Harvey, J. 1996. Intangible assets as determinants of advanced manufacturing technology adoption in SMEs. IEEE Transactions on Engineering Management. (In press.)

——— 1996. Intangible assets as determinants of advanced manufacturing technology adoption in SMEs: toward an evolutionary model. IEEE Transactions on Engineering Management. (In press.)

Lefebvre, L.A.; Mason, R.; Lefebvre, É. 1996. The influence prism in SMEs: the power of CEOs' perceptions on technology policy and its organizational impacts. Management Science. (In press.)

Leonard-Barton, D. 1987. Implementing structured software methodologies: a case of innovation in process technology. Interfaces, 17(3), 6–17.

Leonard-Barton, D.; Deschamps, I. 1988. Managerial influence in the implementation of new technology. Management Science, 34, 1252–1265.

Lincoln, D.J.; Warberg, W.B. 1987. The role of microcomputers in small business marketing. Journal of Small Business Management, 25(2), 8–17.

Litvack, I.A.; Warner, T.N. 1987. Multinationals, advanced manufacturing technologies, and Canadian public policy. Business Quarterly, Summer, 14–19.

Majchzak, A. 1988. The human side of factory automation. Jossey-Bass Publishers, San Francisco, CA, USA.

McFetridge, D.G. 1989. Distinguishing characteristics of users of new manufacturing technologies. Carleton University, Ottawa, ON, Canada.

McWilliams, G. 1994. Mom and Pop go high tech. Business Week, No. 3400, 21 Nov, pp. 82–90.

Meredith, J. 1987. The strategic advantages of the factory of the future. California Management Review, 29, 27–41.

Meyer, A.D.; Goes, J.B. 1988. Organizational assimilation of innovations: a multilevel contextual analysis. Academy of Management Journal, 31(4), 897–923.

Miller, D.; Friesen, P.H. 1982. Innovation in conservative and entrepreneurial firms: two models of strategic momentum. Strategic Management Journal, 3, 1–25.

Miller, J.G.; Roth, A.V. 1988. Manufacturing strategies: executive summary of the 1988 North American manufacturing futures survey. Boston University, Boston, MA, USA. Manufacturing Roundtable Research Report Series.

Montazemi, A.R. 1988. Factors affecting information satisfaction in the context of the small business environment. MIS Quarterly, 12(2), 239–256.

Naik, B.; Chakravarty, A.K. 1992. Strategic acquisition of new manufacturing technology: a review and research framework. International Journal of Production Research, 30(7), 1575–1601.

Nazem, S.M. 1990. Sources of software and levels of satisfaction for small business computer applications. Information and Management, 19, 95–100.

Neo, B.S. 1988. Factors facilitating the use of information technology for competitive advantage: an exploratory study. Information and Management, 15, 191–201.

Nichols, W.; Jones, O. 1994. The introduction of CIM: a strategic analysis. International Journal of Operations and Production Management, 14(8), 4–16.

Nilakanta, S.; Scamell, R.W. 1990. The effects of information sources and communication channels on the diffusion of an innovation in a data base environment. Management Science, 36, 24–40.

Palvia, P., Means, D.B.; Jackson, W.M. 1994. Determinants of computing in very small businesses. Information and Management, 27, 161–174.

Pearson, J.M.; Shim, J.P. 1992. An empirical investigation into decision support environments: findings and considerations. Proceedings, 13th International Conference on Information Systems, Dec, Dallas, TX, USA.

Porter, M. 1980. Competitive strategy. Free Press, New York, NY, USA.

Poutsma, E.; Walravens, A., ed. 1989. Technology and small enterprises: technology autonomy and industrial organization. Delft University Press, The Netherlands.

Powell, P. 1992. Information technology and business strategy: a synthesis of the case for reverse causality. Proceedings, 13th International Conference on Information Systems, Dec, Dallas, TX, USA.

Prekumar, G.; Ramamurthy, K.; Nilakanta, S. 1992. The impact of interorganizational relationships on the adoption and diffusion of interorganizational systems. Proceedings, 13th International Conference on Information Systems, Dec, Dallas, TX, USA.

Rabeau, Y.; Lefebvre, L.A.; Bourgault, M. 1994. Impact de la technologie de l'information sur l'organisation et les stratégies de l'entreprise : revue de littérature et modèles émergents. Centre for Information Technology Innovation, Supply and Services Canada, Ottawa, ON, Canada. C028–1/114.

Raho, L.E.; Belohlav, J.A.; Fiedler, K.D. 1987. Assimilation of new technology into the organization: an assessment of McFarlan and McKenny's model. MIS Quarterly, 11(1), 43–56.

Raymond, L. 1985. Organizational characteristics and MIS success in context of small business. MIS Quarterly, 9(1), 37–52.

Rivazd, S.; Huff, S.L. 1988. Factors of success for end user computing. Communications of the ACM, 31(5), 552–561.

Robertson, T.S.; Gatignon, H. 1987. The diffusion of high technology innovations: a marketing perspective. In Pennings, J.M.; Buitendam, A., ed., New technology as organizational innovation. Ballinger Publishing Company, Cambridge, MA, USA.

Röller, L.H.; Tombak, M.M. 1993. Competition and investment in flexible technologies. Management Science, 39(1), 107–117.

Rothwell, R. 1992. Successful industrial innovation: critical factors for the 1990s. R&D Management, 22(3), 221–239.

Schleich, J.F.; Corney, W.J.; Boe, W.J. 1990. Microcomputer implementation in small business: current status and success factors. Journal of Microcomputer Systems Management, Fall, 2–10.

Schroeder, D.M.; Gopinath, G.; Congden, S.W. 1989. New technology and the small manufacturer: panacea or plague? Journal of Small Business Management, 27, 1–10.

Stair, R.M.; Crittenden, W.F.; Crittenden, V.L. 1989. The use, operation and control of the small business computer. Information and Management, 16(3) 125–130.

Straub, D.W. 1994. The effect of culture on IT diffusion: e–mail and FAX in Japan and the US. Information Systems Research, 5(1), 23–47.

Takac, P.F. 1994. Outsourcing: a key to controlling escalating IT costs? International Journal of Technology Management, 9(2), 139–155.

Thomas, R.; Saren, M.; Ford, D. 1994. Technology assimilation in the firm: managerial perceptions and behaviour. International Journal of Technology Management, 9(2), 227–240.

Torkzadeh, G.; Dwyer, D.J. 1994. A path analytic of determinants of information system usage. Omega, 22(4), 339–348.

US Bureau of the Census. 1993. Manufacturing technology: factors affecting adoption 1991. US Bureau of the Census, Department of Commerce, Washington, DC, USA. Current Industrial Reports, SMT (91)–2.

Venkatraman, N. 1989. Strategic orientation of business enterprises: the construct, dimensionality, and measurement. Management Science, 35, 942–962.

Will, M.M. 1986. The status of microcomputer utilization by selected small businesses. Journal of Computer Information Systems, Fall.

Willcocks, L.; Griffiths, C. 1994. Predicting risk of failure in large-scale information technology projects. Technological Forecasting and Social Change, 47, 205–228.

Winter, S.J.; Gutek, B.A.; Chudoba, K.M. 1992. Misplaced resources? Factors associated with computer literacy among end–users. Proceedings, 13th International Conference on Information Systems, Dec, Dallas, TX, USA.

Woudhuysen, J. 1994. Tailoring IT to the needs of customers. Long Range Planning, 27(3), 33–42.

Wroe, B. 1987. Successful computing in a small business. NCC Publications, Manchester, UK.

Yap, C.S. 1989. Computerisation problems facing small and medium enterprises: the experience of Singapore. Proceedings, 20th Annual Meeting of the Midwest Decision Sciences Institute, 19–21 Apr, Miami University, Miami, FL, USA. pp. 128–134.

Zaheer, A.; Venkatraman, N. 1994. Determinants of electronic integration in the insurance industry: an empirical test.Management Science, 40(5), 549–566.

Zicklin, G. 1987. Numerical control machining and the issue of deskilling, Work and Occupations, 14, 452–466.

Chapter 3. Characteristics of the Decision-making Process as Prime Adoption Factor

Adler, P.S. 1986. New Technologies, new skills. California Management Review, 29(1), 9–28.

———— 1989. CAD/CAM: managerial challenges and research issues. IEEE Transactions on Engineering Management, 36, 202–215.

Amoako, K.; Gyampah; Maffei, M.J. 1989. The adoption of flexible manufacturing systems: strategic considerations. Technovation, 9, 479–491.

Baker, W.H. 1987. Status of information management in small businesses. Journal of Systems Management, 38(4), 10–15.

Bantel, K.A.; Jackson, S.E. 1989. Top management and innovations in banking: does the composition of the top team makes a difference? Strategic Management Journal, 10, special issue, 107–124.

Breakwell, G.M.; Fife-Shaw, C.; Lee, T.; Spencer, J. 1987. Occupational aspirations and attitudes to new technology. Journal of Occupational Psychology, 60, 169–172.

Byod, B.K.; Dess, G.G.; Rasheed, A.M.A. 1993. Divergence between archival and perceptual measures of the environment: causes and consequences. Academy of Management Review, 18(2), 204–226.

Chen, M. 1986. Gender and computers: the beneficial effects of experience on attitudes. Journal of Educational Computing Research, 2(3), 265–282.

Cooper, R.B.; Zmud, R.W. 1990. Information technology implementation research: a technological diffusion approach. Management Science, 36, 123–139.

Davis, F. 1989. Perceived usefulness, perceived ease of use, and user acceptance of information technology. MIS Quarterly, 13(3), 319–340.

Davis, F.; Bagozzi, R.; Warshaw, R. 1989. User acceptance of computer technology: a comparison of two theoretical models. Management Science, 35, 982–1003.

Dean, J.W. 1987. Deciding to innovate. How firms justify advanced technology. Ballinger Publishiung Company, Cambridge, MA, USA.

Dearden, J. 1969. The case against ROI control. Harvard Business Review, May/Jun, 124–135.

Finnie, J. 1988. The role of financial appraisal in decisions to acquire advanced manufacturing technology. Accounting and Business Research, 18(7), 133–139.

Gattiker, U.E. 1990. Technological adoption and organizational adaption: developing a model for human resource management in an international business environment. Monographs in Organizational Behavior and Industrial Relations. Vol. 11: Organizational issues in high technology management. JAI Press Inc., Greenwich, CT, USA. pp. 265–296.

Gattiker, U.E.; Larwood, L., ed. 1988. Computer end–users: the impact of their beliefs on subjective career success. In Managing technological development: strategic and human resources issues. Walter de Gruyter, New York, NY, USA. Ch. 8.

Gattiker, U.E.; Nelligan, T.W. 1988. Computerized offices in Canada and the United States: investigating dispositional similarities and differences. Journal of Organizational Behavior, 9, 77–96.

Gattiker, U.E.; Gutek, B.A.; Berger, D.E. 1988. Office technology and employee attitudes. Social Science Computer Review, 6(3), 327–340.

Gerwin, D.; Kolodny, H. 1992. Management of advanced manufacturing technology — strategy, organization and innovation. John Wiley & Sons Inc., New York, NY, USA.

Gupta, Y. 1988. Organizational issues of flexible manufacturing systems. Technovation, 8(4), 255–268.

Hill, T. 1994. Manufacturing strategy (2nd ed.). Irwin, Boston, MA, USA.

Huff, S.L.; Munro, M.C. 1989. Managing micro proliferation. Journal of Information Systems Management, 72–75.

Kumar, V.; Loo, S. 1988. The adoption of advanced manufacturing technologies: an analysis of the investment decision-making process. Proceedings, POM–ASAC Conference, Halifax, NS, Canada.

Lefebvre, É.; Lefebvre, L.A. 1990. The importance of planning computer acquisitions: the case of small business. Journal of Small Business and Entrepreneurship, 8(3), 56–67.

———— 1992. CEO characteristics and technology adoption in smaller manufacturing firms. Journal of Engineering and Technology Management, 9, 243–277.

———— 1993. Entrepreneurial intensity and the adoption of process innovations: the case of small manufacturing firms. Vol. 2, No. 4. Creativity and innovation management. Basil Blackwell, Oxford, UK. pp. 228–236.

Lefebvre, É.; Lefebvre, L.A.; Roy, M.J. 1995. Technological penetration; organizational learning in SMEs: the cumulative effect. Technovation, 15(8), 511–522.

Lefebvre, L.A.; Harvey, J.; Lefebvre, É. 1991. Technological experience and the technology adoption decisions in small manufacturing firms. R&D Management, 21, 241–249.

Lefebvre, L.A.; Lefebvre, É.; Harvey, J. 1996. Intangible assets as determinants of advanced manufacturing technology adoption in SMEs: toward an evolutionary model. IEEE Transactions on Engineering Management. (In press.)

Lefebvre, L.A.; Mason, R.; Lefebvre, É. 1996. The influence prism in smes: the power of CEOs' perceptions on technology policy and its organizational impacts. Management Science. (In press.)

Martin, C.J. 1989. Information management in the smaller business: the role of the top manager. International Journal of Information Management, 9, 187–197.

Meredith, J. 1987. The strategic advantages of new manufacturing technologies for small firms. Strategic Management Journal, 8, 249–258.

Meredith, J.; Hill, M.M. 1987. Justifying new manufacturing systems: a managerial approach. Sloan Management Review, Summer, 49–61.

Meyer, A.D.; Goes, J.B. 1987. How organizations adopt and implement new technologies. 47th Annual Meeting of the Academy of Management, New Orleans, LA, USA. pp. 175–179.

Moynihan, T. 1990. What chief executives and senior managers want from their IT departments. MIS Quarterly, March, 15–25.

Naik, B.; Chakravarty, A.K. 1992. Strategic acquisition of new manufacturing technology: a review and research framework. International Journal of Production Research, 30(7), 1575–1601.

Nickell, G.S.; Seado, P.C. 1986. The impact of attitudes and experience on small business computer use. American Journal of Small Business, Spring.

Nilakanta, S.; Scamell, R.W. 1990. The effects of information sources and communication channels on the diffusion of an innovation in a data base environment. Management Science, 36, 24–40.

Pennings, J.M. 1987. On the nature of new technology as organizational innovation. *In* Pennings, J.M.; Buitendam, A., ed., New technology as organizational innovation. Ballinger Publishing Company, Cambridge, MA, USA.

Rogers, E.M. 1983. Diffusion of innovations. Free Press, New York, NY, USA.

Roth, A.V.; Miller, J.G. 1988. 1988 North American manufacturing futures survey fact book. Boston University, Boston, MA. Manufacturing Roundtable Research Report Series.

Schroeder, D.M.; Gopinath, G.; Congden, S.W. 1989. New technology and the small manufacturer: panacea or plague? Journal of Small Business Management, 27, 1–10.

Sioukas, A.V. 1995. User involvement for effective customization: an empirical study on voice networks. IEEE Transactions on Engineering Management, 42(1), 39–49.

Son, Y.K. 1992. A Comprehensive bibliography on justification of advanced manufacturing technologies. Engineering Economist, 38, 59–71.

Swann, K.; O'Keefe, W.D. 1990. Advanced manufacturing technology: investment decision process. Part 1. Management Decision, 28(1), 20–31.

———— 1990. Advanced manufacturing technology: investment decision process. Part 2. Management Decision, 28(3), 27–34.

Thacker, R.M. 1989. A new CIM model: a blueprint for the computer-integrated manufacturing enterprise. SME, Dearborn, MI, USA.

Troxler, J.W. 1990. Estimating the cost impact of flexible manufacturing. Journal of Cost Management in Manufacturing, 4(2), 26–32.

Voss, C.A. 1988. Success and failure in advanced manufacturing technology. International Journal of Technology Management, 3, 285–297.

Zammuto, R.F.; O'Connor, E.J. 1992. Gaining advanced manufacturing technologies' benefits: the roles of organization design and culture. Academy of Management Review, 17(4), 701–728.

Chapter 4. Impacts of Adoption

McKenny, J.L. 1995. Waves of change: business evolution through information technology. Harvard Business School Press, Boston, MA, USA.

Relationship between IT and productivity

Alpar, P.; Kim, M. 1990. A comparison of approaches to the measurement of IT value. Proceedings, 22nd Hawaiian International Conference on Systems Sciences, Honolulu, HI, USA.

Baily, M.; Chakrabarti, A. 1988. Electronics and white-collar productivity. *In* Innovation and the productivity crisis. Brookings, Washington, DC, USA.

Barua, A.; Kriebel, C.; Mukhopadhyay, T. 1991. Information technology and business value: an analytic and empirical investigation. University of Texas, Austin, TX, USA. Working Paper.

Bolwijn, P.T.; Kumpe, T. 1990. Manufacturing in the 1990s — productivity, flexibility and innovation. Long Range Planning, 23(4), 44–57.

Brooke, G. 1991. Information technology and productivity: an economic analysis of the effects of product differentiation. University of Minnesota, MN, USA. PhD dissertation.

Brynjolfsson, E. 1993. The productivity paradox of information technology. Communications of the ACM, 36(12), 67–77.

Brynjolfsson, E.; Hitt, L. 1993. Is information systems spending productive? New evidence and new results. International Conference on Information Systems, Orlando, FL, USA.

Cron, W.L.; Sobol, M.G. 1983. The relationship between computerization and performance: a strategy for maximizing the economic benefits of computerization. Journal of Information Management, 171–181.

Fortune. 1993. 13 Dec.

Franke, R.H. 1987. Technological revolution and productivity decline: computer introduction in the financial industry. Technological Forecasting and Social Change, 31.

Harris, S.E.; Katz, J.L. 1989. Predicting organizational performance using information technology managerial control ratios. Proceedings, 22nd Hawaiian International Conference on Systems Sciences, Honolulu, HI, USA.

Harvey, J.; Lefcbvre, É.; Lefebvre, L.A. 1992. Exploring the relationship between productivity problems and technology adoption in small manufacturing firms. IEEE Transactions on Engineering Management, 39(4), 352–358.

Kelley, M.R. 1994. Productivity and information technology: the elusive connection. Management Science, 40(11), 1406–1425.

Lefebvre, É.; Lefebvre, L.A.; Roy, M.J. 1995. Technological penetration and organizational learning in SMEs: the cumulative effect. Technovation, 15(8), 511–522.

Lefebvre, L.A.; Lefebvre, É. 1987. L'Entreprise innovatrice : un regard vers demain. L'Actualité economique — Revue d'analyse économique, 63(1), 53–76.

———— 1988. The impact of information technology on employment and productivity: a survey. National Productivity Review, 7(3), 219–228.

———— 1991. linking new technology adoption to productivity and quality: towards an evolutive model. *In* Sumanth, D., ed., Management productivity frontiers. Industrial Engineering and Management Press, Norcross. pp. 356–361.

Lefebvre, L.A.; Lefebvre, É.; Colin, D. 1991. Process innovation, productivity, and competitiveness. Canadian Journal of Administrative Sciences, 8(1), 19–28.

Lopez-Rodriguez, V.; Pacheco-Espejel, A.A.; Arce-Arnez, J.C. 1995. Motivation, technology and productivity: a proposal to measure its interaction. *In* Productivity and quality management frontiers — V. Industrial Engineering and Management Press, Norcross. pp. 54–65.

Loveman, G.W. 1988. An assessment of the productivity impact on information technologies. MIT Management in the 1990s. Working Paper No. 88–054.

McMillan, C.J. 1987. The automation triangle: new paths to productivity performance. Business Quarterly, 52(2), 61–67.

Morrison, C.J.; Berndt, E.R. 1990. Assessing the productivity of information technology equipment in the U.S. manufacturing industries. National Bureau of Economic Research. Working Paper No. 3582.

Noyelle, T., ed. 1990. Skills, wages and productivity in the service sector. Westview Press, Boulder, CO, USA.

NRC (National Research Council). 1994. Information technology in the service society. NRC, Washington, DC, USA.

Osterman, P. 1986. The Impact of computers on employment of clerks and managers. Industrial and Labor Relations Review, 39, 175–186.

Pacheco-Espejel, A.A. 1993. La Productividad como una espiral de mejora continua. Revista UPIICSA, México, 1(2), 33–40.

Parsons, D.J.; Gottlieb, C.C.; Denny, M. 1990. Productivity and computers in Canadian banking. Department of Economics, University of Toronto, Toronto, ON, Canada. Working Paper No. 9012.

Porter, M. 1980. Competitive strategy. Free Press, New York, NY, USA.

———— 1990. Competitive advantage of nations. Free Press, New York, NY, USA.

Roach, S.S. 1989. America's white-collar productivity dilemma. Manufacturing Engineering, 104.

———— 1991. Services under siege — the restructuring imperative. Harvard Business Review, 82–92.

Rowe, F. 1994. Data network productivity and competitive behavior: the case of the French commercial banks. Technological Forecasting and Social Change, 46, 29–44.

Siegel, D.; Griliches, Z. 1991. Purchased services, outsourcing, computers and productivity in manufacturing. National Bureau of Economic Research. Working Paper No. 3678.

Strassmann, P.A. 1990. The business value of computers. Information Economics Press, New Canaan, CT, USA.

Weill, P. 1990. Do computers pay off? ICIT Press, Washington, DC, USA.

Impacts of IT on key competitive dimensions and performance

Bakos, Y.J. 1987. Dependent variables for the study of firm and industry-level impacts of information technology. Proceedings, 8th International Conference on Information Systems, Dec, Pittsburgh, PA, USA. pp. 10–23.

Bakos, J.Y.; Treacy, M.E. 1986. Information technology and corporate strategy: a research perspective. MIS Quarterly, Jun, 106–119.

Banerjee, S.; Goldhar, D.Y. 1994. Security issues in the EDI environment. International Journal of Operations and Production Management, 14(4), 97–108.

Bergeron, F.; Raymond, L. 1991. Les Avantages de l'EDI. GREPME, Université du Québec à Trois–Rivières, Trois–Rivières, PQ, Canada.

Blois, K.J. 1988. Automated manufacturing creates market opportunities. Journal of General Management, 13(4), 57–73.

Bonaccorsi, A. 1992. On the relationship between firm size and export intensity. Journal of International Business Studies, 23(4), 605–635.

Bort, R.; Schinderle, D.R. 1994. Using EDI to improve the accounts payable department. Heathcare Financial Management, Jan, 78–84.

Boynton, A.C.; Victor, B.; Pine, J.; Eaker, M. 1992. The post-Fordist transformation: information technology and organizational change. Proceedings, 13th International Conference on Information Systems, Dec, Dallas, TX, USA.

Bradley, S.P.; Hausman, J.A.; Nolan, R.L., ed. 1993. Globalization, technology and competion: the fusion of telecommunications in the 1990s. Harvard Business Review.

Brynjolfsson, E.; Malone, T.W.; Gurbaxani, V.; Kambil, A. 1994. Does information technology lead to smaller firms? Management Science, 40(12), 1628–1644.

Buffa, E.S. 1985. Meeting the competitive challenge with manufacturing strategy. National Productivity Review, Spring, 155–169.

Burrows, B. 1994. The power of information: developing the knowledge based organization. Long Range Planning, 27(1), 142–153.

Cale, E.G.; Curley, K.F. 1987. Measuring implementation outcome: beyond success and failure. Information and Management, 13(5), 245–253.

Chen, I.J.; Small, M.H. 1994. Implementing advanced manufacturing technology: an integrated planning model. Omega, 22(1), 91–103.

Clemons, E.K. 1986. Information systems for sustainable competitive advantage. Information and Management, 11(8), 131–136.

——— 1989. Information systems and business strategy: mini-track on current research. Proceedings, 22nd Hawaiian International Conference on Systems Sciences, Honolulu, HI, USA. pp. 181–183.

Clemons, E.K.; Kimbrough, S.O. 1986. Information systems, telecommunications and their effects on industrial organizations. 7th Annual International Conference on Information Systems, Dec, San Diego, CA, USA. pp. 99–108.

Clemons, E.K.; Knez, M. 1988. Competition and cooperation in information systems innovation. Information and Management, 15, 25–35.

Crowston, K.; Treacy, M.E. 1986. Assessing the impact of information technology on enterprise level performance. 7th International Conference on Information Systems, Dec, San Diego, CA, USA. pp. 299–310.

Daly, D.J. 1985. Technology transfer and Canada's competition performance. Toronto Economic Council, Toronto, ON, Canada. Current Issues in Trade and Investment in Service Industries: US–Canadian Perspectives. pp. 304–333.

Davidow, W.H.; Malone, M.S. 1992. The virtual corporation. Harper Collins Publishers, New York, NY, USA.

Davis, D.D. 1986. Integrating technological, manufacturing, marketing and human resource strategies. *In* Davis, D.D. et al., ed., Managing technological innovation. Jossey–Bass Management Series, Jossey–Bass Publishers, San Francisco, CA, USA.

Dean, J.W. 1987. Building the future: the justification process for new technology. *In* Pennings, J.M.; Buitendam, A., ed., New technology as organizational innovation. Ballinger Publishing Company, Cambridge, MA, USA.

Dean, J.W.; Snell, S.A. 1991. Integrated manufacturing and job design: moderating effects of organizational inertia. Academy of Management Journal, 34(4), 776–804.

Defosse, S.F.; Barr, T.D. 1992. Implementing EDI/EGI will reduce time to market. Industrial Engineering, Aug, 30–31.

De Toni, A.; Nassimbeni, G.; Tonchia, S. 1994. The role of information and service in the supply chain. *In* Khalil, T.M.; Bayraktar, B.A., ed., Management of technology IV. Institute of Industrial Engineers. pp. 1286–1293.

Drilhon, G.; Estimé, M.F. 1993. PME : information technologique et compétitivité. L'Observateur de l'OCDE, No. 182 (Jun/Jul), 31–34.

Dwyer, E. 1990. The personal computer in the small business: efficient office practice. NCC, Basil Blackwell, Oxford, UK.

ECC (Economic Council of Canada). 1987. Making technology work. ECC, Ottawa, ON, Canada. Catalogue No. EC22–141/1987E.

El–Najdawi, M.K.; Stylianou, A.C. 1993. Expert support systems: integrating AI technologies. Communications of the ACM, 36(12), 55–103.

Farley, J.U.; Kahn, B.; Lehmann, D.R.; Moore, W.L. 1987. Modelling the choice to automate. Sloan Management Review, Winter, 5–15.

Ferné, G. 1993. Technologies de l'information : les nouveaux enjeux. L'Observateur de l'OCDE, No. 182 (Jun/Jul), 23–26.

Finlay, D. 1993. Create higher profits with voice messaging. The Office, Aug, pp. 32–54.

Finlay, P.N.; MItchell, A.C. 1994. Perceptions of the benefits from the introduction of CASE: an empirical study. MIS Quarterly, Dec, 353–370.

Garsombke, T.W.; Garsombke, D.J. 1989. Strategic implications facing small manufacturers: the linkage between robotization, computerization, automation and performance. Journal of Small Business Management, 27, 34–44.

Gattiker, U.E.; Howg, L. 1990. Information technology and quality of work life: comparing users with non–users. Journal of Business and Psychology, 5(2), 237–260.

Gerwin, D.; Kolodny, H. 1992. Management of advanced manufacturing technology — strategy, organization and innovation. John Wiley & Sons Inc., New York, NY, USA.

Glazer, S. 1993. Before and after: three office makeovers. Inc. Technology Guide, 15(12), 72–79.

Goldhar, J.D.; Jelinek, M. 1985. Computer integrated flexible manufacturing: organizational, economic and strategic implications. Interfaces, 15(3), 94–105.

Grover, V.; Teng, J.T.C.; Fiedler, K.D. 1993. Information technology enabled business process redesign: an integrated planning framework. Omega, 21(4), 433–447.

Harris, S.E.; Katz, J.L. 1989. Predicting organizational performance using information technology managerial control ratios. Proceedings, 22nd Hawaiian International Conference on Systems Sciences, Honolulu, HI, USA.

Harvey, J.; Lefebvre, L.A.; Lefebvre, É. 1991. Technological change and the customer contact paradigm in services. In Khalil, T.M.; Bayraktar, B.A., ed., Management of technology — III. Industrial Engineering and Management Press, Norcross. pp. 898–906.

———— 1993. Technology and the creation of value in services: a conceptual model. Technovation, 13(8), 481–495.

———— 1994. Competing on process capabilities: quick response to loan requests. In Khalil, T.M.; Bayraktar, B.A., ed., Management of technology — IV. Industrial Engineering and Management Press, Norcross. pp. 1258–1266.

Hiltz, S.R.; Turoff, M. 1993. The network nation: human communication via computer. The MIT Press, Cambridge, MA, USA.

Hottenstein, M.P.; Dean, J.W., Jr. 1992. Managing risk in advanced manufacturing technology. California Management Review, 34, 122–126.

Information Week. 1987. The strategic use of information: seizing the competitive edge. May, pp. 26–62.

Ives, B.; Learmonth, G.P. 1984. The information systems as a competitive weapon. Communications of the ACM, 27(12), 1193–1201.

Jelassi, T.; Figon, O. 1994. Competing through EDI at Brun Passot: achievements in France and ambitions for the single European market. MIS Quarterly, Dec, 337–352.

Johnson, K. 1992. Automated labor/attendance system increase billables and improves cash flow. Industrial Engineering, Aug, 36.

Kauffman, R.J.; Weill, P. 1989. An evaluative framework for research on the performance effects of information technology investments. Proceedings, 10th International Conference on Information Systems, Dec, Boston, MA, USA. pp. 377–388.

Keen, P. 1988. Competing in time: using telecommunications for competitive advantage. Ballinger Publishing Company, Cambridge, MA, USA.

Kettinger, W.J.; Grover, V.; Segars, A.H. 1995. Do strategic systems really pay off? Information Systems Management, Winter, 35–43.

Kettinger, W.J.; Grover, V.; Guha, S.; Segars, A.H. 1994. Strategic information systems revisited: a study in sustainability and performance. MIS Quarterly, Mar, 31–58.

King, W.R.; Grover, V.; Hufnagel, E. 1986. Seeking competitive advantage using information-intensive strategies: facilitators and inhibitors. The 1986 NYU Symposium on Strategic Uses of Information Technology, New York, NY, USA.

Kling, R.; Kraemer, K.L.; Allen, J.; Bakos, Y.,; Gurbaxani, V.; King, J. 1992. Information systems in manufacturing coordination: economic and social perspectives. Proceedings, 13th International Conference on Information Systems, Dec, Dallas, TX, USA.

Kutty, P.R.; Raghavan, K.V. 1992. Exploring the benefits that surround automated systems. Industrial Engineering, Aug, 32–43.

Lefebvre, É.; Lefebvre, L.A.; Harvey, J. 1993. Competing internationally through multiple innovative efforts. R&D Management, 23(3), 227–237.

Lefebvre, É.; Lefebvre, L.A.; Préfontaine, L. 1994. Relating technology management capabilities to use of information technology. Computer Society Press, IEEE. pp. 460–468.

Lefebvre, É.; Lefebvre, L.A.; Roy, M.J. 1995. Technological penetration and cumulative benefits in SMEs. Computer Society Press, IEEE.

Lefebvre, É.; Préfontaine, L.; Lefebvre, L.A. 1993. Innovative efforts as predictors of quality, diversity and cost dimensions of manufacturing strategy. In Sumanth, D.J.; Edosomwan, J.A.; Poupart, R.; Sink, D.S., ed., Productivity and quality management frontiers — IV. Industrial Engineering and Management Press, Norcross. pp. 526–534.

Lefebvre, L.A.; Lefebvre, É.; Poupart, R. 1990. Innovativeness and the strategic edge in small firms. National Productivity Review, 9(3), 313–320.

Lind, M.R.; Zmud, R.W.; Fischer, W.A. 1989. Micro-computer adoption: the impact of organizational size and structure. Information and Management, 16(3), 157–162.

Little, E.K.; Wong, L. 1993. Improving the bottom line with imaging technology. Heathcare Financial Management, Jun, 74–79.

Liu, H.Y.; Tarng, M.Y. 1992. Using information technology for successful document management systems. Industrial Engineering, Aug, 34–35.

MacMillan, I.C. 1983. Preemptive strategies, Journal of Business Strategy, Fall, 16–26.

Markus, M.L.; Turner, J.A. 1992. Transforming the organization: reengineering business processes with information technology. Proceedings, 13th International Conference on Information Systems, Dec, Dallas, TX, USA.

Martinson, M.G.; Schindler, F.R. 1995. Organizational visions for technology assimilation: the strategic roads to knowledge-based systems success. IEEE Transactions on Engineering Management, 42(1), 9–18.

McCune, J.C. 1993. The well-automated salesforce. Small Business Reports, 18(8), 29–38.

McFarlan, F.W. 1984. Information technology changes the way you compete. Harvard Business Review, 62(3), 98–103.

McLure, M.; Barnett, P. 1994. EDI provides strategy for laboratory results reporting. Heathcare Financial Management, Jan, 85–89.

Meredith, J. 1987. The strategic advantages of new manufacturing technologies for small firms. Strategic Management Journal, 8, 249–258.

Moynihan, J.J.; Kibat, G. 1994. EDI for human resources saves money and time. Heathcare Financial Management, Jan, 72–77.

Moynihan, J.J.; Norman, K. 1994. CHIN provides vital healthcare linkages. Heathcare Financial Management, Jan, 59–64.

Naik, B.; Chakravarty, A.K. 1992. Strategic acquisition of new manufacturing technology: a review and research framework. International Journal of Production Research, 30(7), 1575–1601.

Neo, B.S. 1988. Factors facilitating the use of information technology for competitive advantage: an exploratory study. Information and Management, 315, 191–201.

O'Brien, J.A. 1993. Management information systems: a managerial end–user perspective (2nd ed.). Irwin, Boston, MA, USA.

OECD (Organization for Economic Co-operation and Development). 1987. Technologies de l'information et perspectives économiques. OECD, Paris, France.

———— 1988. Nouvelles technologies : une stratégie socio-économique pour les années 90. OECD, Paris, France.

———— 1989. Technologies de l'information et les nouveaux domaines de croissance. OECD, Paris, France. PIIC Series, No. 19.

———— 1992. Réseaux d'information et nouvelles technologies de l'information : les possibilités économiques naissantes et leurs conséquences pour les politiques des TI au cours des années 90. OECD, Paris, France.

———— 1993. Les Indicateurs d'utilisation des TI : une base nouvelle pour la formulation de la politique des TI. OECD, Paris, France. PIIC Series, No. 31.

———— 1994. Perspectives des politiques des technologies de l'information. OECD, Paris, France.

Pant, S.; Rattner, L.; Hsu, C. 1994. Manufacturing information integration using a reference model. International Journal of Operations and Production Management, 14(11), 52–72.

Parker, C.M.; McKinney, J. 1993. information technology and small discounters. Discount Merchandiser, 33(5), 124–127.

Parker, R.S.; Doke, E.R.; Acree, A. 1994. Issues in marketing local area network products. Industrial Marketing Management, 23, 71–81.

Parsons, G.L. 1983. Information technology: a new competitive weapon. Sloan Management Review, Fall, 3–14.

Pedersen, M.K. 1990. Strategic information systems in manufacturing industries. Proceedings, 11th International Conference on Information Systems, Dec, Copenhagen, Denmark, pp. 193–204.

Porter, M. 1985. Competitive strategy. Free Press, New York, NY, USA.

Porter, M.E.; Millar, V.E. 1985. How information gives you competitive advantage. Harvard Business Review, Jul/Aug, 149–169.

Primrose, P.L.; Leonard, R. 1985. Evaluating the intangible benefits of flexible manufacturing systems by use of discounted algorithms within a comprehensive computer program. Proceedings of the Institute of Mechanical Engineers, 199, 23–28.

Puri, S.J.; Sashi, C.M. 1994. Anatomy of a complex computer purchase, Industrial Marketing Management, 23, 17–27.

Röller, L.H.; Tombak, M.M. 1993. Competition and investment in flexible technologies. Management Science, 39(1), 107–114.

Samuels, J.; Greenfield, S.; Mpuku, H. 1992. Exporting and the small firm. International Small Business Journal, 10(2), 24–36.

Schroeder, D.M.; Gopinath, G.; Congden, S.W. 1989. New technology and small manufacturer: panacea or plague? Journal of Small Business Management, 27, 1–10.

Sethi, V.; King, W. 1994. Development of measures to assess the extent to which an information technology application provides competitive advantage. Management Science, 40(12), 1601–1611.

Singhal, R.; Fine, C.H.; Meredith, J.R.; Suri, R. 1987. Research and models for automated manufacturing. Interfaces, 17, 5–14.

Snell, S.A.; Dean, J.W., Jr. 1992. Integrated manufacturing and resource management: a human capital perspective. Academy of Management Journal, 35(3), 467–504.

Srinivasan, K.; Kekre, S.; Mukhopadhyay, T. 1994. Impact of electronic data interchange technology on JIT shipments. Management Science, 40(10), 1291–1304.

Strassmann, P.A. 1994. Information technology and organizational effectiveness. Hawaii International Conference on Systems Sciences, Maui, HI, USA.

——— 1985. Information payoff: the transformation of work in the Electronic Age. Free Press, New York, NY, USA.

Sullivan-Trainor, M. 1989. Building competitive advantage by extending information systems. Computerworld, 23(41), SR/19.

Sun, H. 1994. Patterns of organizational changes and technological innovations. International Journal of Technology Management, 9(2), 213–226.

Synnott, W.R. 1987. The information weapon. John Wiley & Sons, New York, NY, USA.

Teng, J.T.C.; Grover, V.; Fiedler, K.D. 1994. Re–designing business processes using information technology. Long Range Planning, 27(1), 95–106.

Thurik, R. 1993. Exports and small business in The Netherlands: presence, potential and performance. International Small Business Journal, 11(3), 49–58.

Utterback, J.M. 1994. Mastering the dynamics of innovation. Harvard Business School Press, Boston, MA, USA.

Vieller, M. 1989. Rude awakening: the rise, fall and struggle for recovery of General Motors. William Morrow, New York, NY, USA.

Vijayaraman, B.S.; Ramakrishna, H.V. 1993. Disaster preparedness of small businesses with micro-computer based information systems. Journal of Systems Management, 44(6), 28–32.

Williams, J.R.; Novak, R.S. 1990. Aligning CIM strategies to different markets. Long Range Planning, 23(1), 126–135.

Wince-Smith, D.L. 1991. Perspectives on U.S. technological competitiveness. Business and the Contemporary World, 4, 20–26.

Wiseman, C. 1988. Strategic information systems. Richard D. Irwin Inc., Homewood, IL, USA.

Wong, B.K.; Chong, J.K.S.; Park, J. 1994. Utilization and benefits of expert systems in manufacturing: a study of large American industrial corporations. International Journal of Operations and Production Management, 14(1), 38–49.

Wright, B. 1994. Health care and privacy law in electronic commerce. Heathcare Financial Management, Jan, 66–70.

Youssef, M.A. 1994. The impact of the intensity level of computer-based technologies on quality. International Journal of Operations and Production Management, 14(4), 4–25.

Impacts of IT on work, training, and employment, and industrial relations

Allen, D.; McGowan, R. 1986. Labour and training issues in advanced technology industrial development. New Technology, Work, and Employment, 1(1), 27–36.

Asher, J.M. 1988. Cost of quality in service industries. International Journal of Quality and Reliability Management, 5(5), 38–46.

Azzone, C.; Mansella, C.; Bertele, U. 1991. Design of performance measures for time-based companies. International Journal of Operations and Production Management, 11(3), 77–85.

Bessant, J.; Grunt, M. 1985. Management and manufacturing innovation in the UK and Germany. Gower Press, Aldershot.

Blumberg, M.; Gerwin, D. 1984. Coping with advanced manufacturing technology. International Institute of Management, Berlin.

Bonaccorsi, A. 1992. On the relationship between firm size and export intensity. Journal of International Business Studies, 33(4), 605–635.

Butera, F.; Della Roca, G. 1986. Technological innovation, organisation of work, and unions. In Jacobi, O.; Jessop, B.; Kastendiek, H.; Regini, M., ed., Technological change, rationalisation and industrial relations. Croom Helm, London, UK.

Campbell, A.; Warner, M. 1987. New technology, innovation and training: a survey of British firms. New Technology, Work, and Employment, 2(2), 86–99.

CEDEFOP (Centre européen pour le développement de la formation professionnele). 1987. Développements actuels de la politique de l'emploi, de l'éducation et de la formation professionnelle au Japon. European Centre for the Development of Vocational Training. CEDEFOP Flash, No. 6/87.

Cressey, P. 1984. The role of the parties concerned in the introduction of new technology. European Foundation for the Improvement of Living and Working Conditions, Dublin, Ireland.

Cyert, R.; Mowery, D., ed. 1987. Technology and employment: innovation and growth in the U.S. economy. National Academy Press, Washington, DC, USA.

Daniel, W. 1987. Workplace industrial relations and technical change. HMSO, Economic and Social Research Council, Policy Studies Institute, ACAS, and Frances Pinter (Publishers) Ltd, London, UK.

Drucker, P.F. 1993. Post-capitalist society. Harper Business, New York, NY, USA.

Gennard, J.; Dunn, S. 1983. The impact of new technology on the structure and organisation of craft unions in the printing industry. British Journal of Industrial Relations, 21(1).

Groebner, D.F.; Merz, M. 1994. The impact of implementing JIT on employees' job attitudes. International Journal of Operations and Production Management, 14(1), 26–37.

ILO (International Labour Organization). 1985. Les Partenaires sociaux face au changement technologique, 1982–85. International Labour Office, Geneva, Switzerland.

Kauffman, R.J.; Weill, P. 1989. An evaluative framework for research on the performance effects of information technology investments. Proceedings, 10th International Conference on Information Systems, Dec, Boston, MA, USA. pp. 377–388.

Kim, D.-J. 1994. Expansion of the information workforce: innovation pull or automation push? Technological Forecasting and Social Change, 46, 51–58.

Kochan, T.; Tamir, B. 1986. Collective bargaining and new technology: some preliminary propositions. Proceedings, International Industrial Relations Association. pp. 201–212.

Landsbury, R. 1986. L'Évolution technologique et les relations professionnelles. Proceedings, International Industrial Relations Association. pp. 1–18.

Lefebvre, É.; Lefebvre, L.A.; Roy, M.J. 1995. Technological penetration and organizational learning in SMEs: the cumulative effect. Technovation, 15(8), 511–522.

Maffei, M.J.; Meredith, J. 1994. The organizational side of flexible manufacturing technology: guidelines for managers. International Journal of Operations and Production Management, 14(8), 17–27.

Mark, J. 1987. Technological change and employment: some results from BLS research. Monthly Labor Review, 110(4), 26–29.

NAS (National Academy of Science). 1986. Human resource practices for implementing advanced manufacturing technology. National Academy Press, Washington, DC, USA.

NEDO (National Economic Development Office). 1985. Advanced manufacturing technology: the impact of new technology on engineering and batch production. Advanced Manufacturing Systems Group, NEDO, London, UK.

OECD (Organization for Economic Co-operation and Development). 1989. Nouvelles technologies : une stratégie socio-économique pour les années 90. OECD, Paris, France.

———— 1991. Ressources humaines et technologies de fabrication avancées. OECD, Paris, france.

Olszewski, P.; Mokhtarian, P. 1994. Telecommuting frequency and impacts for State of California employees. Technological Forecasting and Social Change, 45, 275–286.

Pedersen, M.K. 1990. Strategic information systems in manufacturing industries. Proceedings, 11th International Conference on Information Systems, Dec, Copenhagen, Denmark. pp. 193–204.

Price, R.; Steininger, S. 1987. Trade unions and new technology in West Germany. New Technology, Work, and Employment, 2(2), 100–111.

Rabeau, Y.; Lefebvre, L.A.; Bourgault, M. 1994. Impact de la technologie de l'information sur l'organisation et les stratégies de l'entreprise : revue de littérature et modèles émergents. Centre for Information Technology Innovation, Supply and Services Canada, Ottawa, ON, Canada. C028–1/114.

Schultz-Wild, R. 1987. Work design and work organisation in flexible manufacturing systems. 10th World Congress on Automatic Control, Munich. International Federation of Automatic Control.

Senker, P.; Beesley, M. 1986. The need for skills in the factory of the future. New Technology, Work, and Employment, 1(1), 9–17.

Stalk, G.; Hout, T. 1990. Competing against time. Free Press, New York, NY, USA.

Swann, K.; O'Keefe, W.D. 1990. Advanced manufacturing technology: investment decision process. Part 1. Management Decision, 28(1), 20–31.

———— 1990. Advanced manufacturing technology: investment decision process. Part 2. Management Decision, 28(3), 27–34.

Troxler, J.W. 1990. Estimating the cost impact of flexible manufacturing. Journal of Cost Management in Manufacturing, 4(2), 26–32.

Vickery, G.; Campbell, D. 1989. Advanced manufacturing technology and the organization of work. STI Review, No. 6, 105–146.

Willman, P. 1986. New technology and industrial relations: a review of the literature. UK Department of Employment. Research Paper No. 56.

———— 1987. Technological change, collective bargaining, and industrial efficiency. Oxford University Press, Oxford, UK.

Operational measures of the impacts of IT at the firm level

Atkinson, J. 1985. Flexibility: planning for an uncertain future. Manpower Policy and Practice, 1, 1–9.

Bakos, Y.J.; Treacy, M.E. 1986. Information technology and corporate strategy: a research perspective. MIS Quarterly, Jun, 107–119.

Bartezzaghi, E.; Turco, F. 1989. The impact of just-in-time on production system performance: analytical framework. International Journal of Production Management, 9(8), 40–61.

Bartezzaghi, E.; Turco, F.; Spina, G. 1992. The impact of just-in-time approach on production system performance: a survey of Italian industry. International Journal of Production Management, 12(1), 5–17.

Bourgault, M.; Lefebvre, É.; Lefebvre, L.A. 1995. Productivity and performance in the new corporation: the need for a new perspective. *In* Productivity and quality management frontiers. Industrial Engineering and Management Press, Norcross. pp. 341–351.

Brill, P.H.; Mandelbaum, M. 1990. Measurement of adaptability and flexibility in production systems. European Journal of Operation Research, 49, 325–332.

Browne, J.; Dubois, D.; Rathmill, K.; Sethi, P.S.; Stecke, K.E. 1984. Classification of flexible manufacturing system. The FMS Magazine, Apr, pp. 114–117.

Buzacott, J.A. 1982. The fundamental principles of flexibility in manufacturing systems. Proceedings, 1st International Conference on Flexible Manufacturing Systems, Brighton, UK. pp. 13–22.

Campbell, J.P.; Campbell, R.J. 1988. Productivity in organizations. Jossey-Bass Publishers, San Francisco, CA, USA.

Carlsson, B. 1989. Technology and competitiveness: the micro–macro links. Case Western Reserve University. Working Paper.

Carter, M.F. 1986. Designing flexibility into automated manufacturing systems. Proceedings, 2nd ORSA–TIMS Conference of FMS.

Clemmer, J. 1990. Firing on all cylinders: the service/quality system for high powered performance. Macmillan of Canada, Toronto, ON, Canada.

De Meyer, A.; Ferdows, K. 1990. Influence of manufacturing improvement programmes on performance. International Journal of Production Management, 10(2), 120–131.

Donleavy, G.D. 1994. Evaluating the potential of office robotics. Long Range Planning, 27(2), 119–127.

Drucker, P.F. 1993. Post-capitalist society. Harper Business, New York, NY, USA.

Garvin, D.A. 1988. Managing quality: the strategic and competitive advantage. Free Press, New York, NY, USA.

Gerwin, D. 1987. An agenda for research on the flexibility process. International Journal of Operations and Production Management, 7(1), 38–49.

Hall, R.W.; Johnson, H.T.; Turney, P.B.B. 1990. Measuring up: charting pathways to manufacturing excellence. Richard D. Irwin Inc., Homewood, IL, USA.

Harrington, H.J. 1991. Business process improvement: the breakthrough strategy for total quality, productivity and competitiveness. McGraw–Hill, New York, NY, USA.

Juran, J.M. 1988. Juran on planning for quality (2nd ed.). Free Press, New York, NY, USA. pp. 70–83.

Juran, J.M.; Grynia, F.M., Jr. 1980. Quality planning and analysis: from product development through use (2nd ed.). McGraw–Hill, New York, NY, USA.

Mairesse, J.; Mohnen, P. 1990. Recherche–développement et productivité : un survol de la littérature économetrique. Économie et Statistique, Nos. 237–238, 99–108.

Maskell, B.H. 1991. Performance measurement for world class manufacturing: a model for American companies. Productivity Press, Cambridge MA, USA.

Meredith, J. 1988. New justification approaches for CIM. Journal of Cost Management in Manufacturing, 15–20.

Miller, J.G.; Roth, A.V. 1988. Manufacturing strategies: executive summary of the 1988 North American manufacturing futures survey. Boston University, Boston, MA, USA. Manufacturing Roundtable Research Report Series.

Mizuno, S. 1988. Company-wide total quality control. Nordica International Limited, Hong Kong.

Noble, J.L. 1989. Techniques for cost justifying CIM. Journal of Business Strategy, Jan/Feb, 44–49.

———— 1990. A new approach for justifying computer-integrated manufacturing. Journal of Cost Management in Manufacturing, 3(4), 14–19.

Noori, H. 1990. Managing the dynamic of new technology. Prentice-Hall, Englewood Cliffs, NJ, USA.

Oakland, J.S.; Wynne, R.M. 1991. Efficiency of U.K. engineering production management systems. International Journal of Operations and Production Management, 11(2), 14–26.

Paltridge, S. 1993. Evaluer la performance des Télécom. L'Observateur de l'OCDE, No. 182 (Jun/Jul), 35–38.

Pasewack, W.R. 1991. The evolution of quality control costs in U.S. manufacturing. Journal of Cost Management in Manufacturing, Apr, 46–52.

Primrose, P.L. 1990. Selecting and evaluating cost-effective MRP and MRPII. International Journal of Production and Management, 10(1), 51–66.

Primrose, P.L.; Leonard, R. 1986. The financial and economic application of advanced manufacturing technology. Proceedings of the Institute of Manufacturing Engineers, 200(B1), 27–31.

Pyoun, Y.S.; Byoung, K. C. 1994. Quantifying the flexibility value in automated manufacturing systems. Journal of Manufacturing Systems, 13(2), 108–118.

Quarante, D. 1984. Éléments de design industriel. Maloine S.A. éditeur, Paris, France.

Sapienza, H.J.; Smith, K.G.; Gannon, M.J. 1988. Using subjective evaluation of organizational performance in small business research. American Journal of Small Business, 12, 43–53.

Sethi, A.K.; Sethi, S.P. 1990. Flexibility in manufacturing: a survey. International Journal of FMS, 2, 289–328.

Sethi, V.; King, W. 1994. Development of Measures to Assess the Extent to Which an Information Technology Application Provides Competitive Advantage, Management Science, vol.40, no.12, p.1601–1611.

Sink, D.S.; Tuttle, T.C.; De Vries, S.J. 1984. Productivity measurement and evaluation: what is available? National Productivity Review, 3(3), 265–287.

Statistics Canada. 1992. Aggregate productivity measures, system of national accounts 1990–1991. Minister of Industry, Science and Technology, Ottawa, ON, Canada. Catalogue No. 15–204E.

Sumanth, D.J. 1984. Productivity engineering and management. McGraw-Hill, New York, NY, USA.

Tayntor, C.B. 1994. Managing end–user computing: partners in excellence. Information Systems Management, Winter, 81–83.

Thurow, L.C. 1992. Head to head. William Morrow and Company, New York, NY, USA.

Wilson, R.L. 1994. An improved goal-oriented method for measuring productivity. International Journal of Operations and Production Management, 14(1), 50–59.

About the Institution

The International Development Research Centre (IDRC) is committed to building a sustainable and equitable world. IDRC funds developing-world researchers, thus enabling the people of the South to find their own solutions to their own problems. IDRC also maintains information networks and forges linkages that allow Canadians and their developing-world partners to benefit equally from a global sharing of knowledge. Through its actions, IDRC is helping others to help themselves.

About the Publisher

IDRC Books publishes research results and scholarly studies on global and regional issues related to sustainable and equitable development. As a specialist in development literature, IDRC Books contributes to the body of knowledge on these issues to further the cause of global understanding and equity. IDRC publications are sold through its head office in Ottawa, Canada, as well as by IDRC's agents and distributors around the world.

About the Authors

Élisabeth Lefebvre holds MBA and MSc degrees from the University of Ottawa, as well as a PhD from the University of Montréal. She is currently an associate professor in the Department of Mathematics and Industrial Engineering at the École Polytechique, Montréal. Dr Lefebvre's research focuses on technological innovation and the strategic management of technology.

Louis A. Lefebvre holds an MBA from the University of Ottawa and a PhD in business adminstration from the University of Montréal. He is currently a professor in the Department of Mathematics and Industrial Engineering at the École Polytechnique, Montréal — where he also directs the graduate program on technology management — and President Elect of the International Association for Management of Technology. Professor Lefebvre's research focuses on technology adoption in small and medium-sized enterprises, firm innovativeness, and strategic alliances.